AI绘画教程

Midjourney关键词灵感手册

白无常 李伊宁 编著

人民邮电出版社
北京

图书在版编目（CIP）数据

AI绘画教程 : Midjourney关键词灵感手册 / 白无常，李伊宁编著. -- 北京 : 人民邮电出版社，2024.1
ISBN 978-7-115-62904-3

Ⅰ. ①A… Ⅱ. ①白… ②李… Ⅲ. ①图像处理软件－教材 Ⅳ. ①TP391.413

中国国家版本馆CIP数据核字(2023)第200358号

内 容 提 要

本书通过图文结合的形式，生动展示了 AI 艺术作品的创作过程和技巧。第 1 部分为软件入门，以简明易懂的语言讲解使用 AI 技术创作艺术作品的方法。第 2 部分为灵感图集，收集了 70 多位 AI 美术馆同学的多样创作灵感图，这些艺术作品展示了 AI 在艺术创作中的无限潜力。该书旨在帮助读者更好地理解 AIGC 艺术，引领他们走进这一领域，为热爱艺术与科技的读者提供学习资料，帮助他们创造出令人惊叹的艺术作品。

本书附赠部分作品电子稿和对应关键词，方便读者边学边练，举一反三。

本书适合对 AIGC 技术感兴趣的读者，特别是插画、设计、美术等相关领域的从业者、学生阅读学习。

◆ 编　　著　白无常　李伊宁
责任编辑　王　冉
责任印制　马振武
◆ 人民邮电出版社出版发行　　北京市丰台区成寿寺路 11 号
邮编　100164　　电子邮件　315@ptpress.com.cn
网址　https://www.ptpress.com.cn
北京尚唐印刷包装有限公司印刷
◆ 开本：700×1000　1/16
印张：10.25　　2024 年 1 月第 1 版
字数：197 千字　　2024 年 1 月北京第 1 次印刷

定价：59.80 元

读者服务热线：(010)81055410　印装质量热线：(010)81055316
反盗版热线：(010)81055315
广告经营许可证：京东市监广登字 20170147 号

前言

亲爱的读者:

欢迎翻开《AI 绘画教程:Midjourney 关键词灵感手册》。

很高兴能与你们一起遨游充满创意和惊喜的 AI 艺术世界。作为科技与想象力的新生代，AIGC 技术充满魅力且具备无限可能性，它像一位有趣又强大的朋友，等待着我们去发掘、认识及了解。通过与它的互动，我们可以探索更多元化的艺术形式和创作模式。它可以帮助我们突破传统的限制，创造出独特且令人惊叹的艺术作品。

通过深入研究探索 AI 的各种可能性，并加以千百次的实践总结，我们认真编撰了一系列教程，并收集了大量灵感图，希望能帮助读者轻松理解和应用这些技巧。

在研究探索中，我们对艺术与科技的交融深深着迷，也逐渐明白了 AI 艺术创作的深刻意义。AI 相当于工业革命时期的照相机和汽车，它淘汰的并不是某种职业，而是不会使用和不愿去了解它的人。科技时代的高速发展，让人类创造了一个又一个奇迹，这些奇迹的出现归功于那些肯努力耕耘、踏实积累并敢想敢为的创新者。AI 的出现并不能取代什么，我们只有正确地使用并引导它，才能全面发挥它的作用，换句话说，懂得使用它的人，才能予其生命，赋其意义，让其闪耀，继而照亮彼此的未来。

无论你是艺术爱好者、学生，还是专业艺术家，我相信，本书都将是你宝贵的学习资料和创作指南。

愿你创作愉快，艺术之门将永远为你敞开!

编者

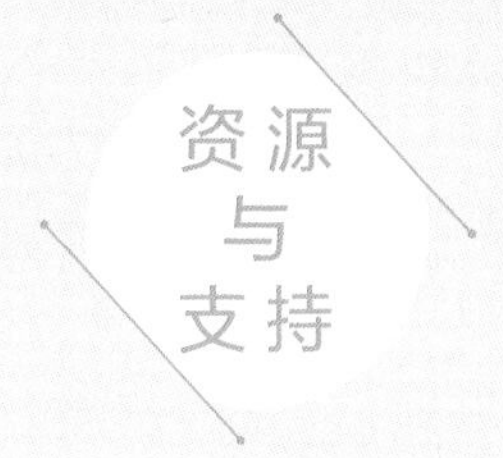

本书由“数艺设”出品，“数艺设”社区平台（www.shuyishe.com）为您提供后续服务。

随书资源：部分作品电子稿和对应关键词。

扫码获取学习资源

（提示：微信扫描二维码关注公众号后，输入51页左下角的5位数字，获得资源获取帮助。）

“数艺设”社区平台 为艺术设计从业者提供专业的教育产品。

与我们联系

我们的联系邮箱是szys@ptpress.com.cn。如果您对本书有任何疑问或建议，请您发邮件给我们，并请在邮件标题中注明本书书名及ISBN，以便我们更高效地做出反馈。

如果您有兴趣出版图书、录制教学课程，或者参与技术审校等工作，可以发邮件给我们。如果学校、培训机构或企业想批量购买本书或“数艺设”出版的其他图书，也可以发邮件联系我们。

关于“数艺设”

人民邮电出版社有限公司旗下品牌“数艺设”，专注于专业艺术设计类图书出版，为艺术设计从业者提供专业的图书、视频电子书、课程等教育产品。出版领域涉及平面、三维、影视、摄影与后期等数字艺术门类，字体设计、品牌设计、色彩设计等设计理论与应用门类，UI设计、电商设计、新媒体设计、游戏设计、交互设计、原型设计等互联网设计门类，环艺设计手绘、插画设计手绘、工业设计手绘等设计手绘门类。更多服务请访问“数艺设”社区平台www.shuyishe.com。我们将提供及时、准确、专业的学习服务。

目录

01 软件入门

02 灵感图集

软件入门

下面主要讲解软件的基础知识，包含账号注册方法、基本设置、关键词描述、垫图方法等，以简明易懂的图文步骤及效果图对比形式帮助读者掌握软件的基本操作方法和技巧。

注册 Midjourney 账号

1. 进入 Midjourney 官网。
2. 按照提示创建账号，进行注册。

注册 Discord 账号

1 进入 Discord 官网，按照提示注册 Discord 账号。

2 授权登录。单击登录页面左下角“需要新的账号？”右侧的“注册”，填写邮箱信息，通过邮件进行验证。

3 输入手机号码，通过验证码进行验证。

Discord 注册注意事项：

（1）需要电子邮箱和手机号双验证，但只能绑定其中一个。

（2）邮箱验证受网络条件限制，若报错，继续等待即可。

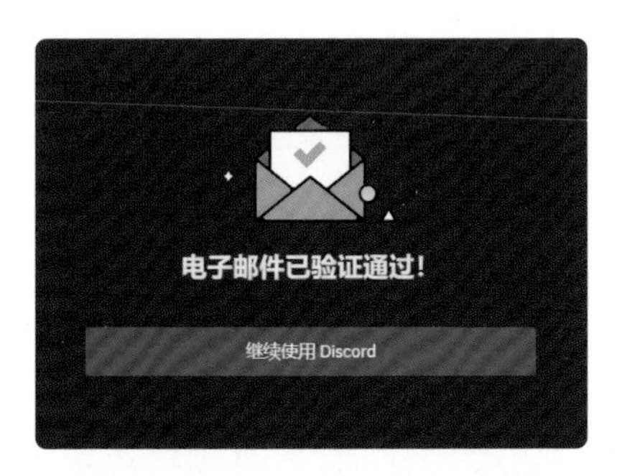

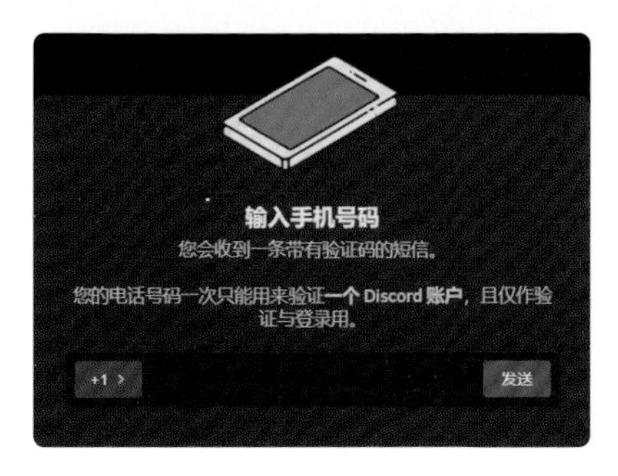

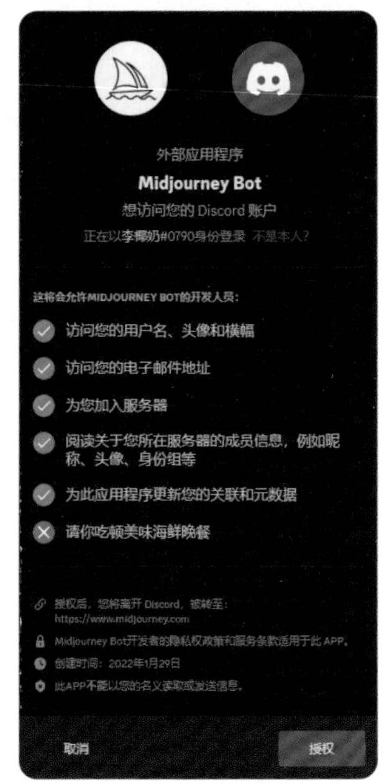

添加生成机器人

1 进入 Midjourney 官网，单击 Join the beta 按钮。

2 接受 Midjourney 的邀请。

3 进入 Discord 界面。

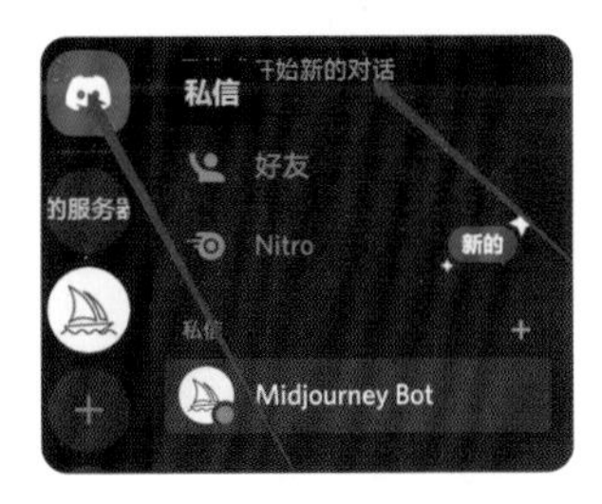

4 选择一个机器人，如 Midjourney Bot，单击 Midjourney Bot。

5 单击“添加至服务器”按钮，完成机器人的添加。

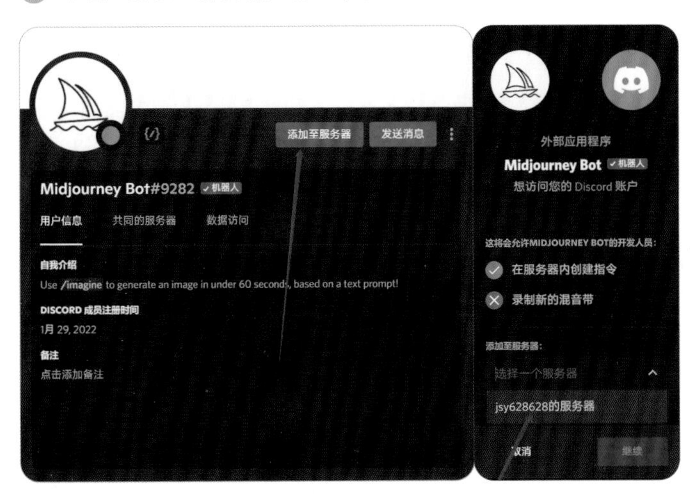

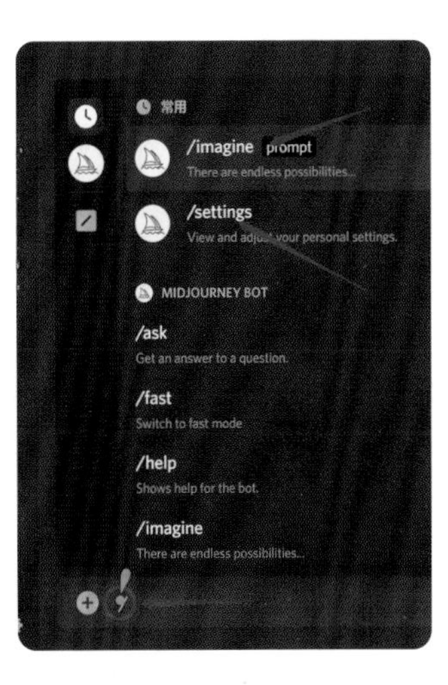

充值

1. 打开 Midjourney 官网，进入个人账号界面。
2. 单击左侧菜单栏中的 Manage Sub（管理子系统）按钮，进入购买页面。

推荐选择 30 美元 / 月的服务，支持登录多台设备。

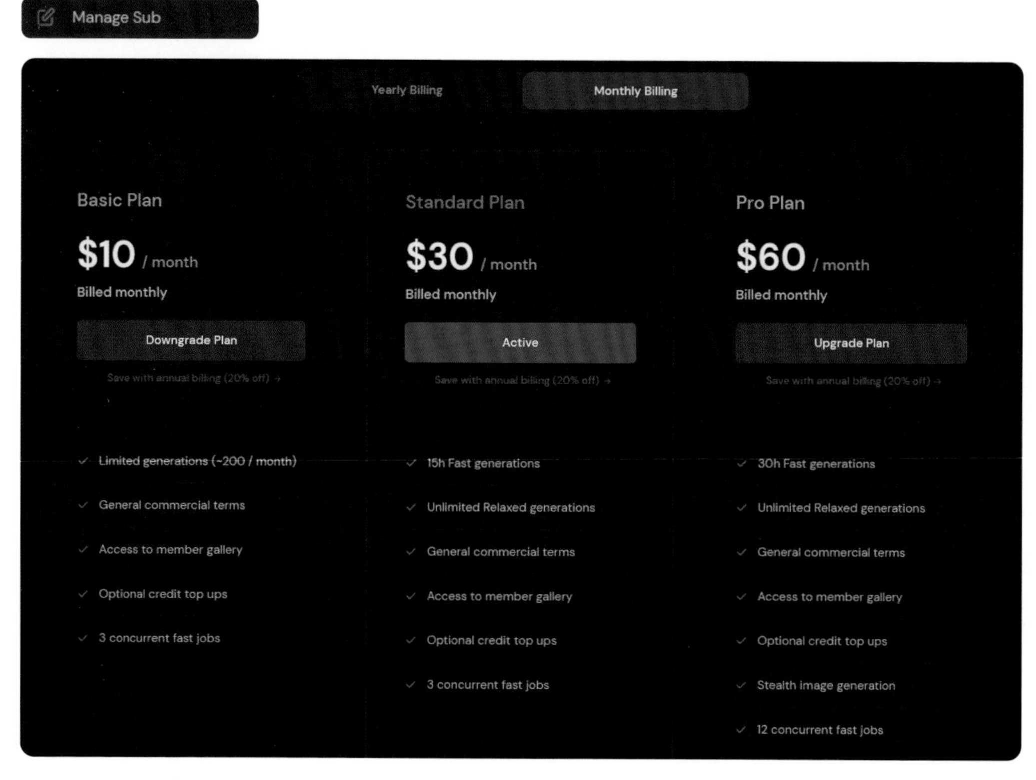

10 美元 / 月　　30 美元 / 月　　60 美元 / 月

设置 (/settings)

进入 Midjourney 的出图界面后，首先要先了解 /settings 指令，可用于基础设置。

① 在对话框中输入 /settings。

② 下方出现相关设置按钮。

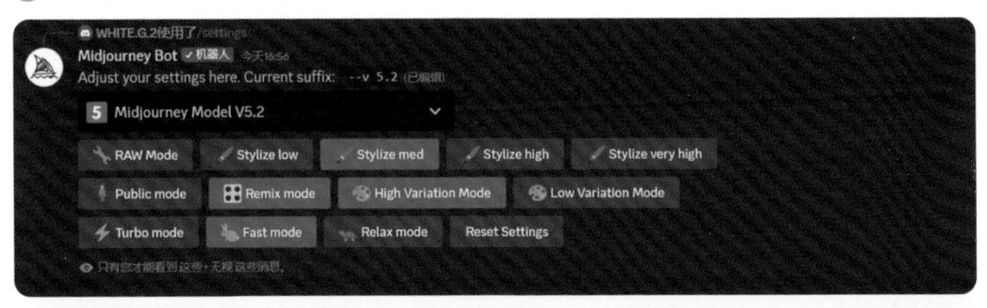

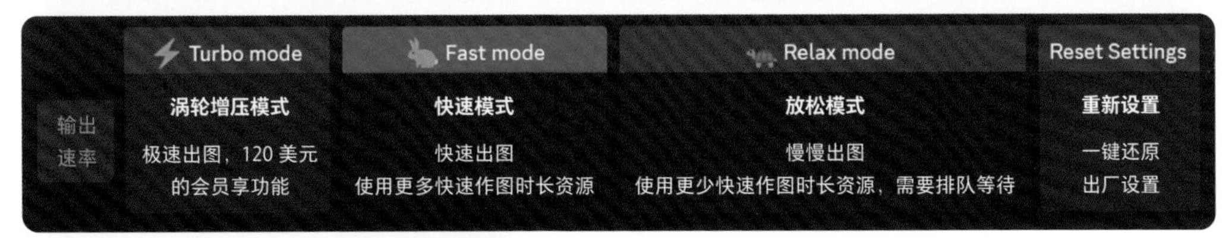

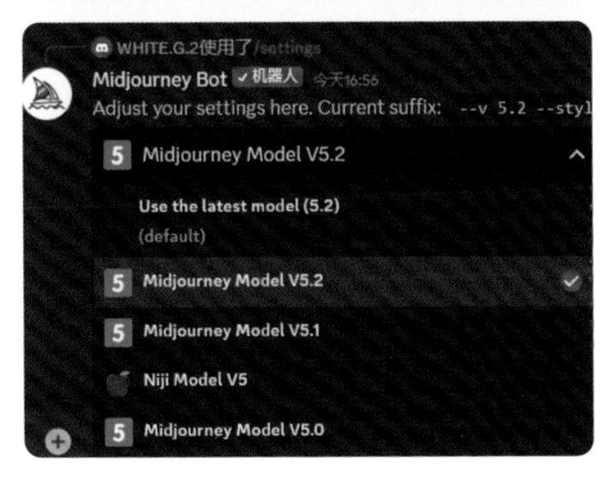

单击⌄按钮，可以展开预设的出图模式，这是 Midjourney 不断更新的出图机器人。

不同出图模式的区别

不同出图模式的出图风格不同，下面用同样的关键词在不同的出图模式下进行测试，对比每个出图模式的画面效果，了解其演变过程。

举例：

一个爱冒险的 8 岁可爱男孩，他喜欢探索和了解世界。

V1

V2

V3

V4

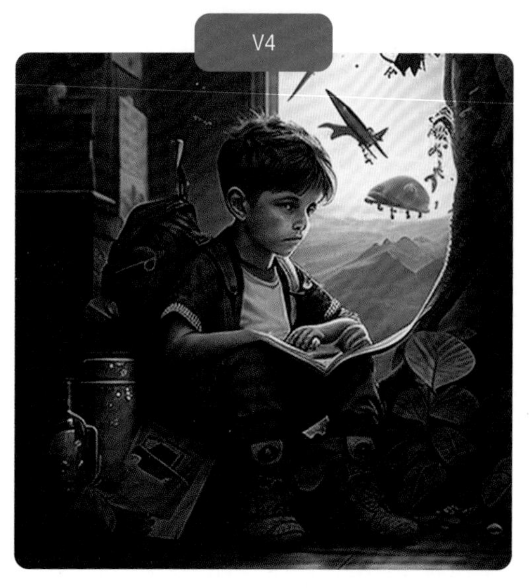

V5

V1 ~ V3 版本生成的图片都比较具有“实验性”，整体简陋不成形，在这一阶段人们使用 AI 绘画更多是在探索新科技带来的可能性。

V4 版本生成的图片整体组织能力、想象力和美术效果都有了质的飞跃，“AI 绘画热”正式从这一版本开始。

V5 版本生成的图片写实性大幅提高，从这一版本开始，AI 绘画不再局限于小范围的探索，而是真正成为可以应用于设计、摄影、影视等多个领域的超强工具。

之后，软件不断更新，推出了 V5.1、V5.2 等版本。下面对比各个版本的出图效果。

写实风格
可模拟真实的肌理、光照
细节丰富

更艺术化
更具想象力
用更简洁的关键词生成细节更丰富的画面

V5.1 版本的 RAW 模式
更艺术化、更具想象力
更注重回归写实风格效果（RAW 即回归）

V5.2 是 V5.1 的加强版
此版本可延展画面尺寸
可以用更简洁的关键词生成细节更丰富的画面

Zoom Out 1.5x 模式下
画面视野扩大 1.5 倍
软件会自动填充周围的空白

Zoom Out 2x 模式下
画面视野扩大 2 倍
软件会自动填充周围的空白，可多次设置

Niji V4 和 Niji V5 生成的图都偏二次元风格，它们就像 V4 和 V5 的升级版一样，Niji V5 出图细节更多，明暗关系处理得更好，图片也更风格化。

可以看到，用同样的关键词出图，只有 Stylize 程度不同（出图时会自动体现在 --s 数值里），Stylize very high 生成的图片艺术化程度更高。

Niji 模型版本是 Midjourney 和 Spellbrush 合作研发的，更擅长生成动漫风格的图片，尤其是长动态和动作镜头，以及以角色为中心的画面。

Niji 机器人添加方式：

① 单击 Midjourney 界面顶部的对话框。

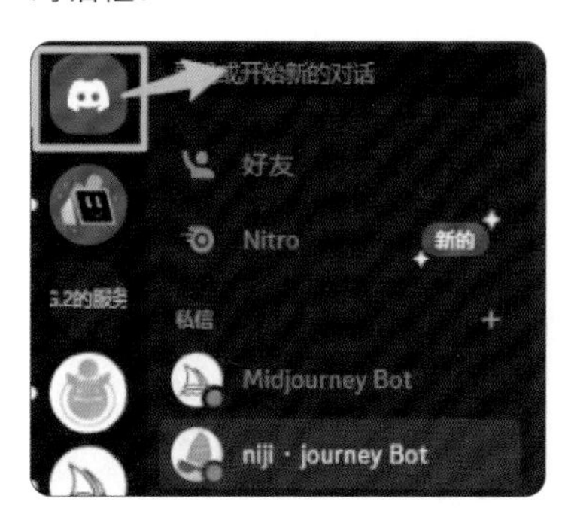

② 输入 niji，选择 niji · journey，添加频道。

③ 左侧新增 Niji 频道，关注资讯。

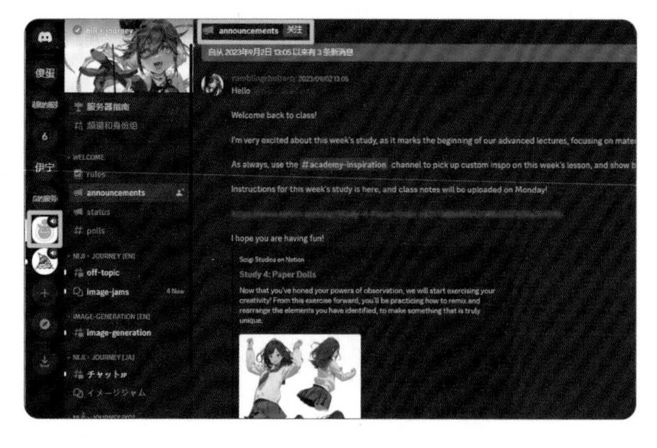

④ 回到原服务器，输入 /settings，选择绿色帆船图标。

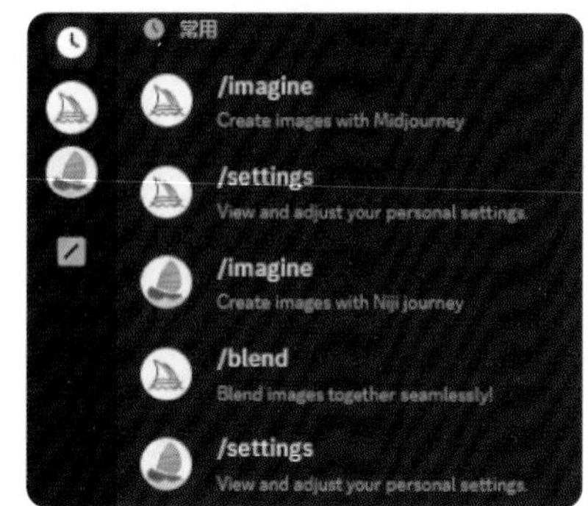

⑤ 弹出 Niji V5 模型版本的 4 个新的设定模型。

不同 Niji 机器人出图的区别

Default Style	Expressive Style	Cute Style	Scenic Style
默认模式	**表现力模式**	**可爱模式**	**风景模式**
二次元爱好者喜爱的模式，与 V5 相同	三维表现力更强 人物表现得更有张力	舒缓、治愈的风格 装饰性更强	背景更加丰富 有充满想象力的远景

举例：

一个爱冒险的 8 岁可爱男孩和鸟，他喜欢探索和了解世界。

想象 (/imagine)

前面讲解了不同模式的出图风格和效果，在设置好想要的出图模式后，就可以输入关键词了。

在对话框中输入 /imagine，按 Enter 键，在对话框内输入画面关键词。

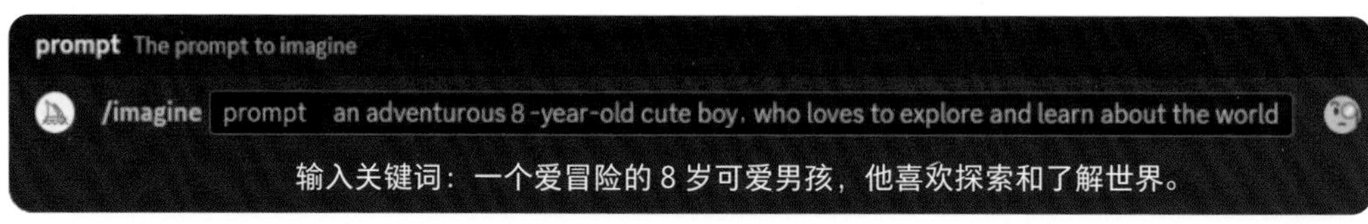

关键词描述

关键词的描述是有规律的，常用的关键词组合公式如下：
主体描述 + 环境场景 + 艺术风格 + 媒介材料 + 摄像机视角 + 精度定义。

主体描述
一句话表达想要的画面，包括修饰词、主体、场景和动作等。

环境场景
主体所处的环境，包括光线、氛围、色系等。

艺术风格
写出想要的设计风格或想要参考的艺术家、漫画作品、影视作品或艺术网站的名字等。

媒介材料
以什么媒介呈现，可以是油画、水彩、摄影、手稿、雕塑、陶瓷、布料、黏土或石头等。

摄像机视角
想要的画面角度，可以是特写、两点透视图、广角镜头、正视图或全身照等。

精度定义
想要的作品质量，可以是高品质、含复杂细节、高分辨率、高清、2K、4K 或给出具体的出图比例。

如果想要的是 3D 效果的画面，需要写出 3D 软件、渲染器，要求的材质、光照及渲染风格等。

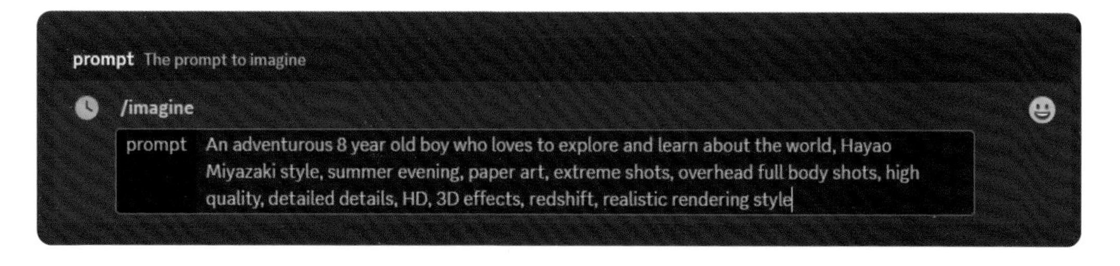

举例：

一个爱冒险的8岁可爱男孩，他喜欢探索和了解世界，在森林里，皮克斯动画风格，C4D，Octane渲染，半身像，黏土雕塑材质，电影照明，高质量，丰富细节，高清。

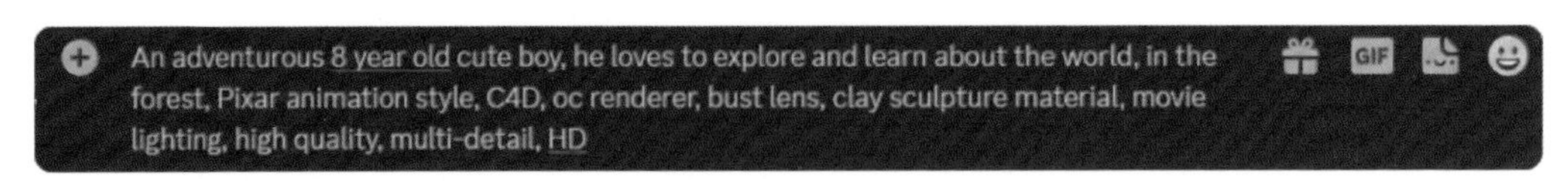

生成 4 幅图片。

其他基本指令

/imagine：根据提示生成图片。

/info：查看账户状态及当前排队或者正在运行的作业状态等信息。

/settings：查看和调整 Midjourney Bot 的设置。

/describe：上传一幅图片，根据图片生成文字描述，供参考使用。

/blend：将 2 ～ 5 幅图融合在一起生成新图像，不支持文本提示。

/help：显示有关 Midjourney Bot 的基本信息和提示。

/fast：切换到快速模式，出图速度快。

/relax：切换到放松模式，出图速度慢。

/stealth：隐身模式，作品不会放在公开空间。

/public：公开模式，默认模式。作品可以供其他人参考。

/show：使用图像 ID 重新生成作品。ID 可以在个人空间通过执行“图片”→“更多”→“COPY”→“Job ID”进行获取。也可以在图像文件名和图片链接中获得，Job ID 后面的字符串即为图片 ID。/show 指令只能重新生成自己的图片。

/prefer option set：创建或管理自定义选项。

/prefer option list：查看当前的自定义选项。

/prefer suffix：设置要添加到关键词末尾的固定后缀。

/prefer remix：开启合成模式，在该模式下每次生成图片之后都可以重新修改关键词与参数。

/subscribe：跳转到订阅服务。

/ask：得到问题的答案。

“傻瓜式”混图

前面讲解了关键词的写作技巧，但是如果不知道怎么描述，又想看看两幅图的混合效果怎么办？这时就会用到 /blend 混图模式。

① 输入 /blend，按 Enter 键，上传图片（扩展名为 .png 或 .jpg）并按 Enter 键，等待 1 ~ 2 分钟。

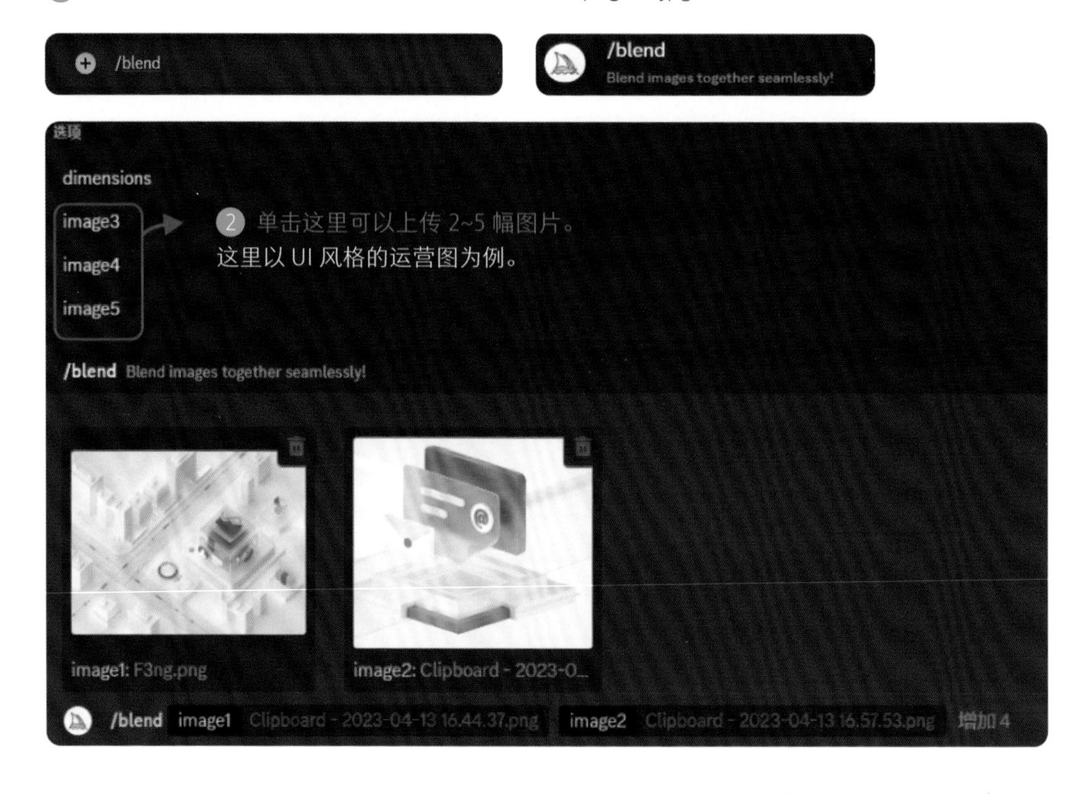

③ 出图结果如左图所示。

· 可以选择同类型、同色系的两幅图，生成的图会相对稳定，比较符合预期。

· 也可以选择不同色系的两幅图进行融合，一幅作为主体，另一幅作为场景，生成的图会让人有意外惊喜。

· 不同模式融合生成出来的图会有风格上的差异，可以多进行尝试以找到符合自己需求的。

垫图的方法

常常听人说垫图很厉害，那怎么垫图呢？很简单，下面来试试。

① 上传文件（将计算机内的图片拖曳到软件中），按 Enter 键，可以一次上传多幅图片。

② 在浏览器中打开图片，按快捷键 Ctrl+C 复制图片的链接地址。如果要用网络上的图片，直接单击鼠标右键选择“复制图片地址”即可。

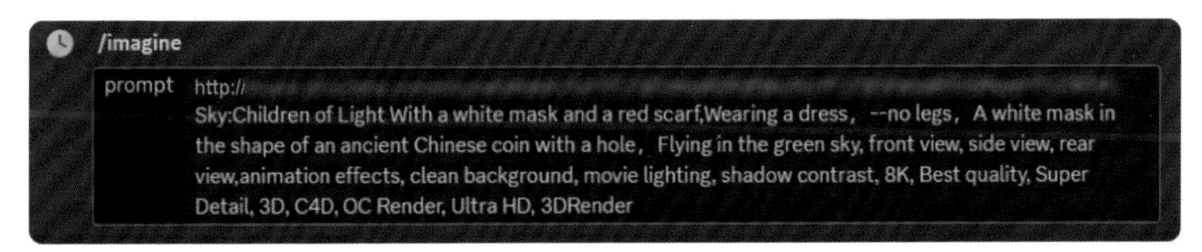

③ 输入 /imagine+ 图片地址（可以是多个图片的地址，中间用空格隔开）+ 空格 + 关键词，然后按 Enter 键。

图生词的方法

当你看到一幅很好看的图片，却有一点“词穷”，不知道怎么描述时，可以使用 Midjourney 的图生词功能来查看描述图片的关键词。

① 输入 /describe，上传图片，按 Enter 键。

② 弹出 4 组不同的描述词，任意选择一组。

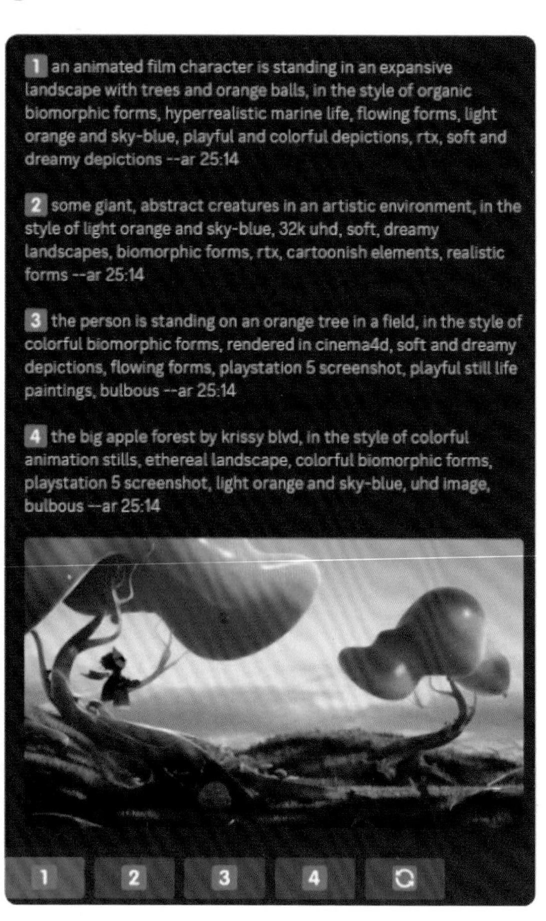

③ 在 Remix 模式下会弹出一个对话框，提示可以修改部分关键词，此处先不修改。

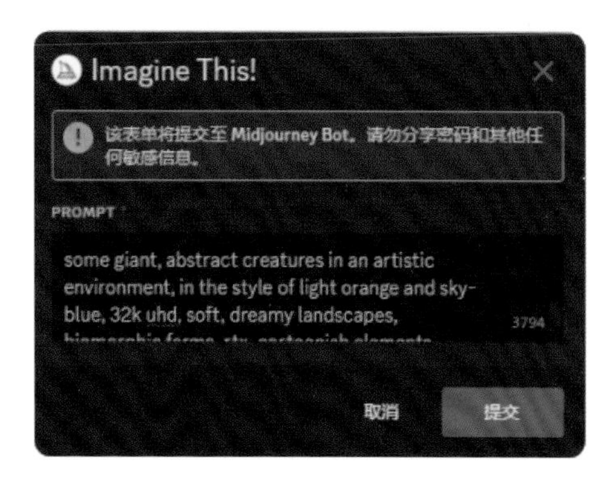

④ 用软件的图生词功能得到的关键词来生成图片，效果如下。

进阶玩法：混合应用

对前面生成的图不太满意怎么办？下面来试一下进阶玩法。

① 改变出图版本，看哪一个生成的效果更好，通过对比可以看到，V5 版本的生成效果更好。

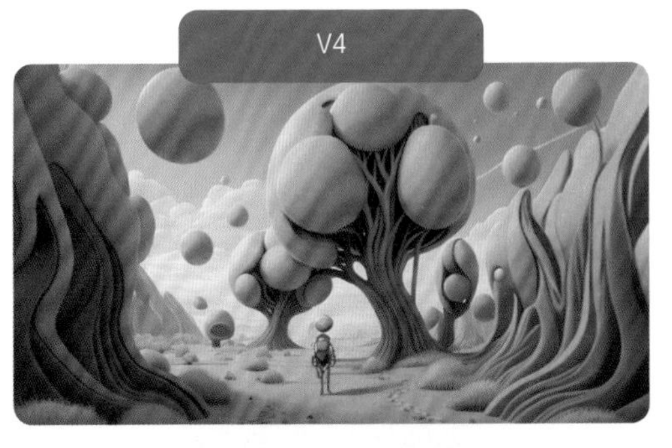

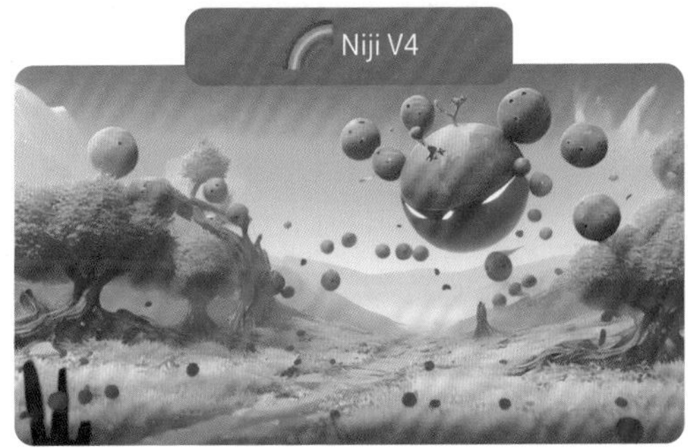

② 确定出图版本后，在图生词的基础垫上原图。

③ 垫图后生成的图片效果与原图很接近，在此基础上稍微修改关键词，或稍微提高原图的权重，然后就可以批量出画风稳定的图片了！

图片权重（--iw 数值）

简单来说，权重就是让谁更重要。

1. --iw 数值可用来控制垫图相对于文字描述的权重。

   V5 版本的数值区间为 0.5 ~ 2，数值越接近 2，垫图对生成图的影响越大。不加 --iw 后缀时，出图时的默认设置为 --iw 0.25，即默认垫图的权重为 25%。

--iw 0.5

--iw 2

2. 输入提示词和后缀时需要注意空格的位置，空格输错、多输或少输都会使整条指令无效。

   正确的写法为：垫图的图片链接 + 空格 + 关键词 + 空格 + --iw+ 空格 + 数值。

关键词权重（:: 数值）

给关键词增加权重，是对关键词的比重进行重新分配的方式，也就是在关键词后加 :: 数值。

正确的书写格式为关键词 +:: 数值，“::”是英文状态下的双冒号且中间无空格。

下面以关键词女孩和老虎 /The girl and the tiger 为例，分别给女孩和老虎不同的权重，对比生成图片的效果。

The girl and the tiger（默认权重）

采用默认权重时，女孩和老虎重要程度相近

The girl and the tiger::2

设置老虎 ::2 时，老虎占比是女孩的两倍

The girl::1 and the tiger

The girl and the tiger::1

通过对比可以看到，给谁加权重，谁就更重要，在画面中的占比就更大。

关键词负权重（::– 数值）和顺序

在关键词后添加 ::– 数值，可以减少或去掉画面里的指定元素，也可以用 --no+ 关键词来实现此操作。

The girl::–5 and the tiger

The girl::–1 and the tiger

关键词的前后顺序也会在一定程度上影响初始的权重分配。

- 下面左图女孩的权重数值为 2，画面中几乎看不到老虎了。
- 下面中图老虎的权重数值为 10，画面中老虎和女孩对比差异不大。
- 将老虎的关键词前移，再将老虎的权重数值设置为 2，将女孩的关键词放后面，这时画面中就只剩老虎了。

The girl::2 and the tiger

The girl and the tiger::10

The tiger::2 and the girl

由此可以得出两个结论：

（1）关键词本身的顺序会影响初始权重的分配。

（2）提示词的有效权重数值为 –1 ～ 2。

尾缀参数含义

--aspect / --ar：长宽比，调整图片的长宽比。

--chaos / --c：混沌值，改变结果的多样性。

--no+ 关键词：负提示，与使用负权重来控制不想出现的元素的作用一样。

--quality/--q：质量，渲染要求的质量越高，渲染时间越长。

--sameseed/--seed：种子，为生成的每幅图片随机生成编号。

--stop：停止，在流程中途完成作业，后面接的数字是百分比。

--stylize/--s：风格，代表美学风格的程度。

--uplight：轻型升频器，优化放大后的图片更平滑、细节更少（V4 版本下）。

--upbeta：详细升频器，优化放大后的图片细节更多、更不光滑（V4 版本下）。

--video：视频，录制生成图像的过程。

--tile：无缝贴图。

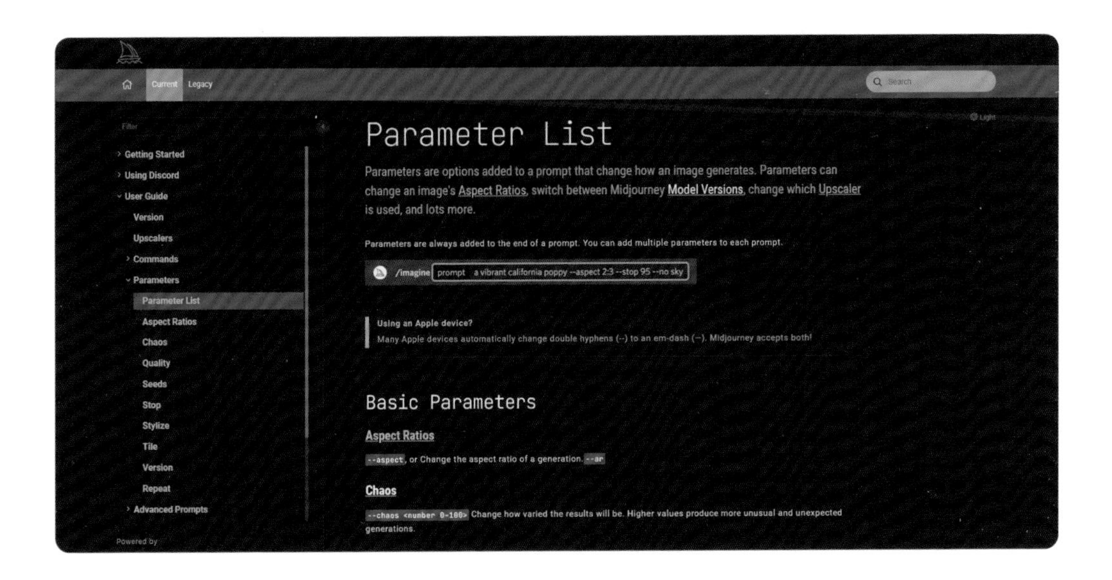

长宽比 (--aspect/--ar)

· 使用 Midjourney 生成图片时的默认长宽比为 1：1，Niji 版本默认的出图长宽比为 3：2。

· 如果想调整图片的长宽比为 16：9，可以通过在关键词后增加尾缀 --aspect 16:9 或 --ar 16:9 来实现。

操作比较简单，但是有几个要注意的点。

（1）在英文和数字中间要加空格。

（2）数字必须使用整数，不能是 --ar 1.5:2。

（3）长宽比会影响图片上内容的形状和整体构图，从图片可以看出，相同的关键词（如鬼龙吃人），不同的长宽比生成的画面效果差异很大。

（4）不同的出图模型支持的最大长宽比不同。

V5	V4c（default）	V4a / V4b	V3	Test / TestP	Niji
任何	1：2~2：1	仅 1：1、2：3 或 3：2	5：2~2：5	3：2~2：3	1：2~2：1

混沌值 (--chaos/--c)

混沌值影响图片的变化程度。

- 设置的混沌值不同，生成的图片效果差异较大。值越低，结果越可靠。
- --chaos 的数值区间是 0~100，默认值为 0。
- 正确的写法是 --chaos 数值或 --c 数值。

举例：

穿着民族服装的卡通女孩背影。

--chaos 0

--chaos 的值越低，生成的结果越可靠、重复性越强

--chaos 100

--chaos 的值越高，生成的图片风格越多样、重复性越低

渲染质量（--quality/--q）

此尾缀代表你要花费多少渲染时间来获取相应渲染质量的图片。

· 默认值为 1，值越高，渲染时间越长。

· 默认的渲染质量值为 0.25、0.5（Half quality）、1（Base quality）和 2x（High quality）。

· 可以输入数值，也可以直接单击相应按钮进行设置。

· 渲染质量的数值不是越高越好，有时越低的数值反倒可以生成效果更好的图片。

· 数值越低，越适合生成具有抽象外观的图片；数值越高，生成的图片细节越丰富。具体设置多少，取决于想要生成什么类型的图片。

举例：

穿着民族服装的卡通女孩背影。

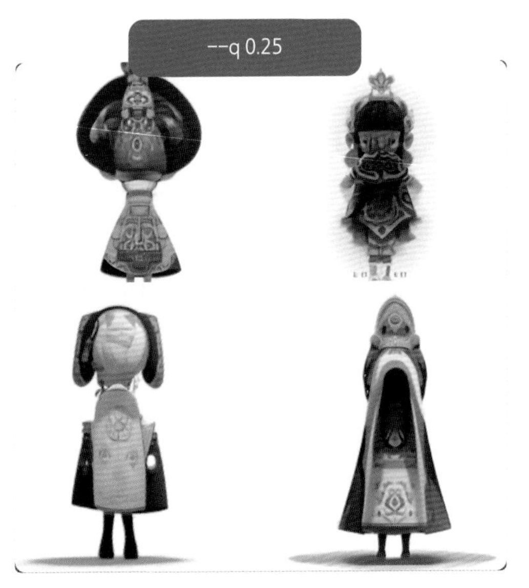

出图快，细节少
出图速度比默认的快 4 倍

默认设置 Base quality
兼顾细节和出图速度

种子（--sameseed/--seed）

添加此尾缀可为生成的每幅图片随机编号，也可以用参数来指定编号。使用相同的种子编号和关键词将产生相似构图、色彩和细节的初始图片网格（仅影响初始图片网格）。

正确的输入方式为 --sameseed 或 --seed。

如何获得种子编号？

1 单击右侧的表情图标，然后单击信封图标，给自己发一封私信。

2 界面左侧出现一个小红点，为未读消息提示，单击打开，即可复制种子编号。

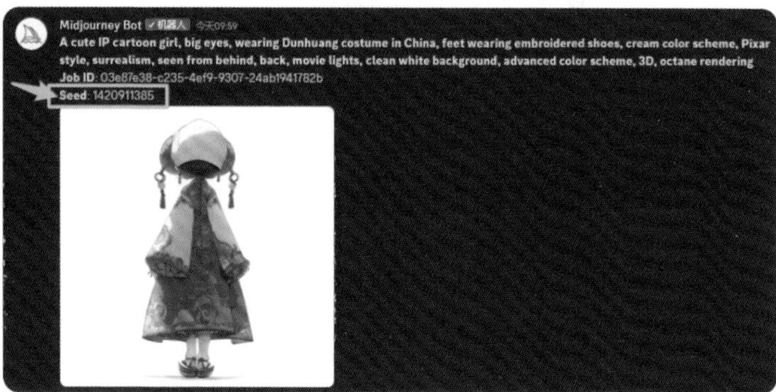

3 返回 Discord 服务器，粘贴种子编号，格式为关键词 + 空格 + --seed+ 空格 + 数字。

修改少量的关键词后，会得到一幅构图、配色都与原图高度相似的图片。

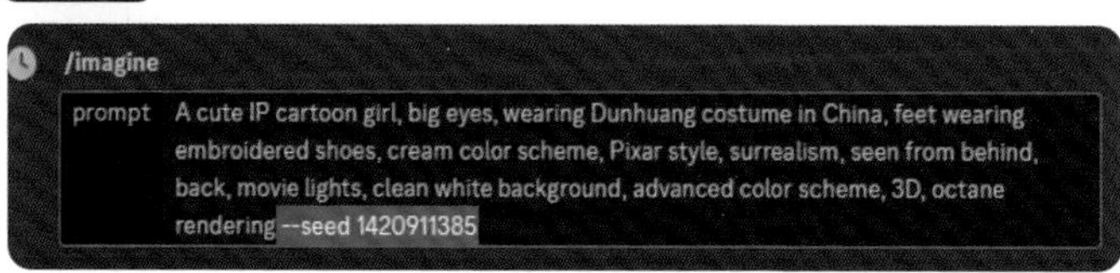

停止（--stop）和视频（--video）

添加 --stop 尾缀可在生成图片的过程中中止作业，数值越低，生成的图片效果越模糊、越不详细。

· --stop 的数值为 10~100，默认值为 100。

Midjourney Bot ✓机器人 今天11:04
A cute IP cartoon girl, wearing Dunhuang costume in China, feet wearing embroidered shoes, cream color scheme, Pixar style, surrealism, seen from behind, back, movie lights, clean white background, advanced color scheme, 3D, octane rendering --stop 50 --niji 5 --q 2 -

· 正确写法为关键词+空格+--stop+空格+数值。

· 在生成一幅图像时，会出现一个生成进度（0 ~ 100%），中途停止将生成更柔和、细节更少的图像。

添加 --video 尾缀可保存出图过程的视频。单击信封图标，可以对软件完成的工作做出反馈，让机器人将视频的链接发给你。

· --video 仅适用于 V1~V3 版本，不适用于其他版本。

· 获取视频链接的方法与获得种子编号的方法类似。

1. 添加 --video 到关键词的末尾。
2. 单击界面右侧的表情图标，找到并单击信封图标，做出反馈。
3. 单击界面左侧的未读消息提示❶图标，再单击页面中的链接即可在浏览器中查看视频。单击鼠标右键或长按链接可以下载视频。

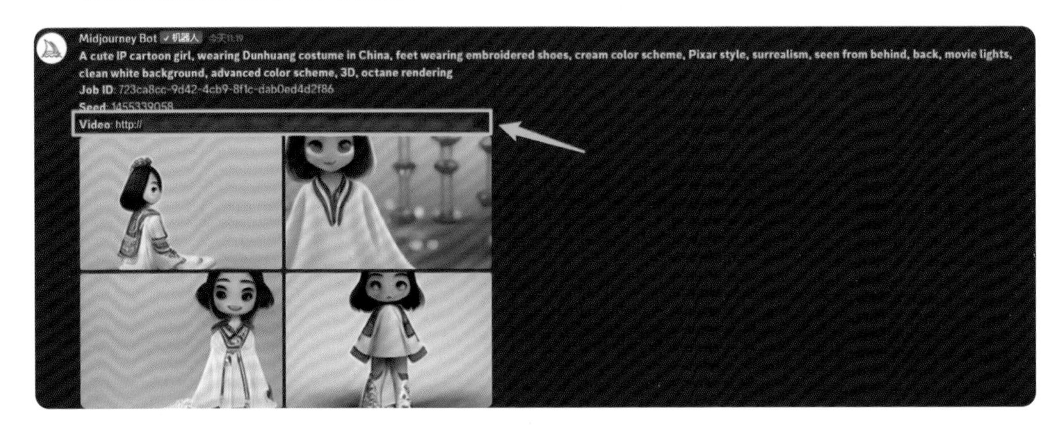

风格（--stylize/--s）

经过训练的 Midjourney，可以生成具有艺术配色、恰当构图和形式的图片。

此参数会影响风格化的强度。用低数值生成的图片效果与关键词描述的内容匹配度更高，但艺术性较差，用高数值生成的图片艺术性更强，但与关键词描述的内容相差较大。

· --stylize 或 --s 的默认值为 100，不同模型版本的风格化数值范围不同。

模型版本	V5	V4	V3	Test / Testp	Niji
默认值	100	100	2500	2500	NA
风格化范围	0~1000	0~1000	625~60000	1250~5000	NA

· 正确写法为 --stylize+ 空格 + 数字或 --s+ 空格 + 数字。

举例：（Niji V5）

一个可爱的 IP 卡通女孩，穿着敦煌风格的服装和绣花鞋，奶油风的配色，皮克斯风格，超现实主义，后视图，电影灯光，干净的白色背景，高级配色，3D，Octane 渲染。

数值低的更符合描述，但放大看，作品很生硬，光影细节处理得不好

数值高的不完全符合描述，也并不是 3D 渲染风格，但整体完成度更高，风格化更强

升频器（--upbeta / --uplight）

· V1~V4 版本和 Niji 版本生成的图片尺寸比较小（512px × 512px），单击 U1 按钮、U2 按钮、U3 按钮和 U4 按钮可将图片放大至 1024px × 1024px，这个优化过程叫作升频。

· V5 版本和 Niji 5 版本生成的图片本身尺寸就是 1024px × 1024px，单击 U1 按钮、U2 按钮、U3 按钮和 U4 按钮，只是把 4 幅图分开，并不会再放大。

模型版本	起始网格大小（px）	V4 默认升频器（px）	细节高档（px）	轻装上档（px）	测试版高档（px）	动漫高档（px）	最大高档（px）
V 4（default）	约 512 × 512	约 1024 × 1024	约 1024 × 1024	约 1024 × 1024	约 2048 × 2048	约 1024 × 1024	-
V5	-	-	-	-	-	-	-
V1~V3	约 256 × 256	约 1024 × 1024	约 1024 × 1024	约 1024 × 1024	约 1024 × 1024	约 1024 × 1024	约 1664 × 1664
Niji	约 512 × 512	约 1024 × 1024	约 1024 × 1024	约 1024 × 1024	约 2048 × 2048	约 1024 × 1024	-
Niji 5	约 1024 × 1024	-	-	-	-	-	-

· 在优化 V4 生成的某一幅图时，下方会出现新的选项按钮，即 Beta 升频器和 Light 升频器，这两个选项生成的图像效果略有不同。

Make Variations　Light Upscale Redo　Beta Upscale Redo　Web
Favorite

优化后原图（V4）

优化后的图（注意这两个升频器目前只有 V4 能用）

--upbeta

图像更光滑、细节处理得更少，适合生成卡通角色图像

--uplight

细节更多，增加了质感和纹理，适合生成写实风格图像

无缝贴图（--tile）

使用该指令生成的图可为织物、壁纸和纹理创作无缝图案。

· 使用方法很简单，将尾缀 --tile 放在描述词的最后即可。

· 目前 V1、V2、V3、V5 和 Niji5 版本都支持 --tile，Niji 和 V4 版本不支持。

举例：

敦煌纹样 --tile。

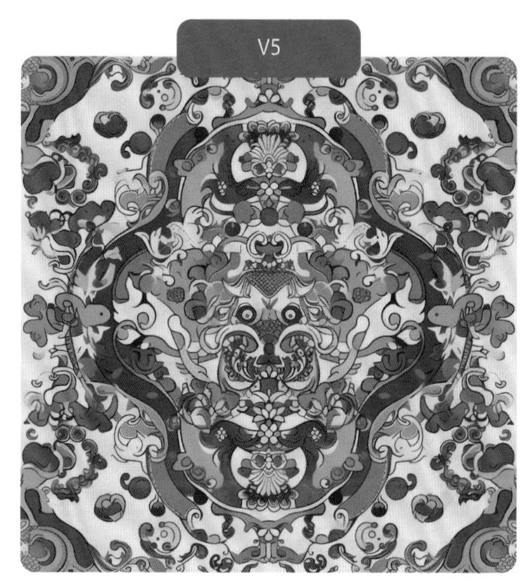

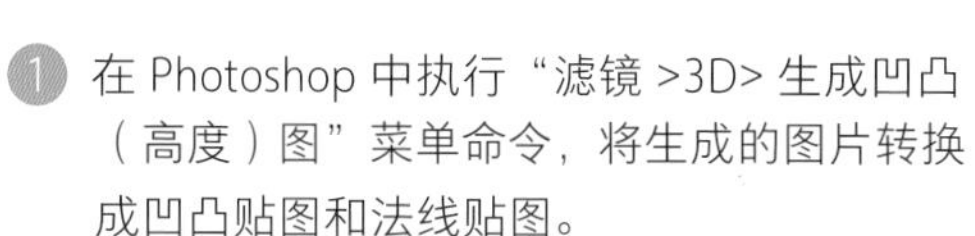

① 在 Photoshop 中执行“滤镜 >3D> 生成凹凸（高度）图”菜单命令，将生成的图片转换成凹凸贴图和法线贴图。

② 将贴图应用到布料上，渲染出接近真实布料的效果。

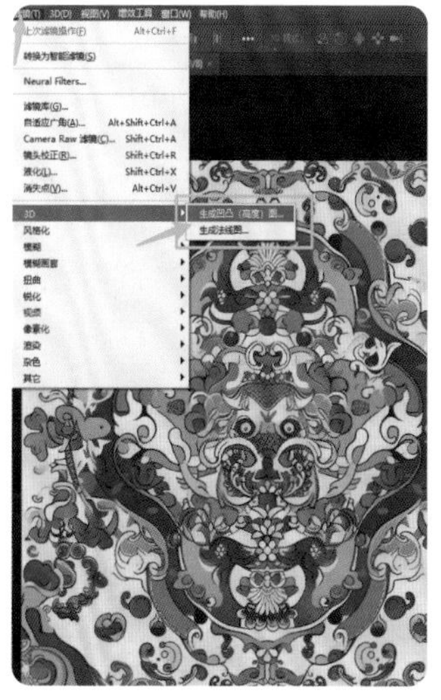

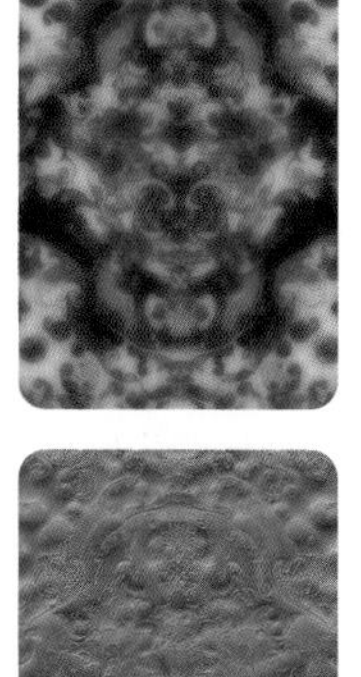

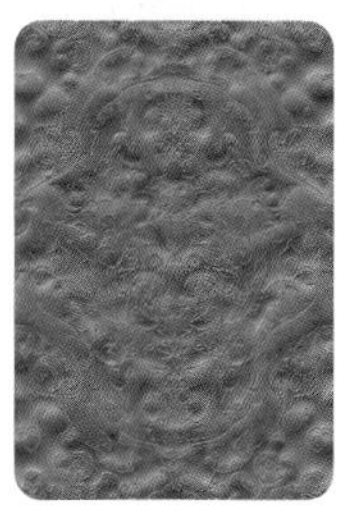

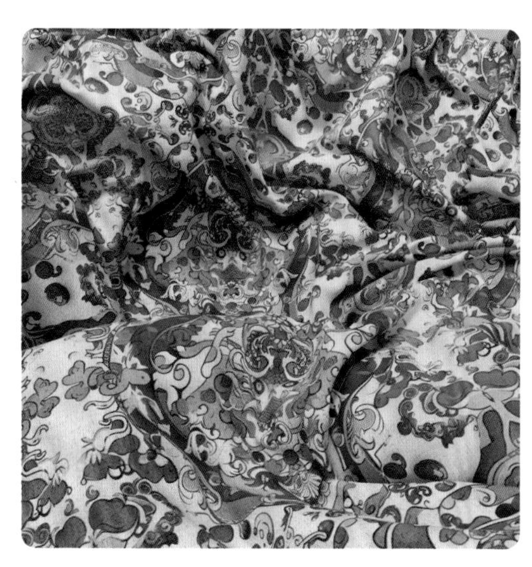

灵感图集

下面展示以食物、配饰、包装等为创作灵感或主题的 AI 绘画作品，并提供生成作品的关键词，以便读者更好地理解 AIGC 技术，读者可在书中展示的作品基础上举一反三，创作出更优秀的作品。

食物

A041－求己

关键词：

from a top-down perspective, exquisite traditional Chinese pastries, the ingredients of traditional Chinese pastries are distributed on the background of traditional shading. the image of the overall atmosphere style. the style will be a photograph, neatly arranged with a high resolution camera and appropriate settings to capture the details of each component, a blue-assisted packaging design, a style of historical reproduction, dreamlike illustration, action comics, precise surrealism, product shooting, medium panorama, deep sculpting, richly detailed, complex and varied textured backgrounds --ar 3:4 --v 5

俯视视角，精致的中国传统糕点，中国传统糕点的配料分布在传统的底纹上。形象的整体风格大气。样式将是一张照片，用一台高分辨率的相机和适当的设置来捕捉排列整齐的每个细节，蓝色辅助包装设计，历史再现风格，梦幻插画，动作漫画，精确的超现实主义，产品拍摄，中全景，深度雕刻，丰富的细节，复杂多样的纹理背景 --ar 3:4 --v 5

A041 - 求己

关键词：

from a top-down perspective, exquisite traditional Chinese pastries, the ingredients of traditional Chinese pastries are distributed on the background of traditional shading. the image of the overall atmosphere style. the style will be a photograph, neatly arranged with a high resolution camera and appropriate settings to capture the details of each ingredient, packaging design, style in historical reproduction, dreamlike illustration, action comics, precise surrealism, product shooting, medium panorama, deep sculpting, richly detailed, complex and varied textured backgrounds --v 5

俯视视角，精致的中国传统糕点，中国传统糕点的配料分布在传统的底纹上。形象的整体风格大气。样式将是一张照片，用一台高分辨率的相机和适当的设置来捕捉排列整齐的每个配料的细节，包装设计，历史再现风格，梦幻插画，动作漫画，精确的超现实主义，产品拍摄，中全景，深度雕刻，丰富的细节，复杂多样的纹理背景 --v 5

S883 - 贝利

关键词：

future jingxiu art, style of still life in the kitchen, Chinese food, cyanine, aerial view, realistic photograph, hyperdetailed, oriental inspired themes, circular shapes, orientalist images --ar 3:4 --v 5 --s 250

未来景秀艺术，厨房静物风格，中餐，花青花，鸟瞰图，写实照片，超多细节，东方风格的主题，圆形，东方主义的图像 --ar 3:4 --v 5 --s 250

U927－路志彤

关键词：

carved by tofu, jiangnan landscape, under water, vegetable garnish, the picture is shrouded in moving smoke --ar 3:4 --q 2 --v 5

豆腐雕刻，江南山水，水下，蔬菜配饰，烟雾缭绕 --ar 3:4 --q 2 --v 5

Q794 - baaacccc

关键词：

Chinese style blue tray bowl with butterflies and pink jelly on top, in the style of use of traditional techniques, sandara tang, atey ghailan, nostalgic imagery, poured resin, traditional essence, light maroon, photography, realistic style --ar 3:4 --q 2 --v 5

中式的蓝色托盘上有蝴蝶和粉红色的果冻，采用传统技法风格，桑达拉汤，阿泰 · 盖兰，怀旧意象，浇上树脂，传统精华，浅栗色，摄影，写实风格 --ar 3:4 --q 2 --v 5

配饰

D159－晓娜

关键词：

Chinese painter's painting, art nouveau illustration style, dark beige and sky blue, full of details, siya oum, ancient Chinese noble depiction, delicate beadwork, digital printing, fine clothing details, profile, folk art inspiration, dark beige and deep sea sapphire --v 5

中国画家的绘画作品，新艺术风格的插画，深米色和天蓝色，充满细节，Siya Oum，中国古代贵族刻画，精致珠饰，数字印刷，精细的服装细节，轮廓，民间艺术灵感，深米色和深海蓝宝石色 --v 5

R827 - 小柒

关键词：

ancient Chinese headdress and jewelry, hairpin, hairpin, bracelet, necklace, jade, gold, silver and jade material, exquisite texture, warm color, light green, light background, top-down perspective, focal length 50, halo 1.8, iso500, award-winning photographic works --s 1000 --niji 5 --style expressive

中国古代的头饰和首饰，发簪、发钗、手镯、项链、玉器、金、银镶玉制品，纹理精致，色彩温暖，浅绿色，浅色背景，俯视视角，焦距 50，光圈 1.8，ISO500，获奖摄影作品 --s 1000 --niji 5 --style expressive

ancient Chinese beauty, wearing light green and white and gray dress, with headdress and jewelry, hairpin, bracelet, necklace, gold, silver and jade material, exquisite texture, warm color, delicate face, elegant posture, bust, light background, focal length 50, halo 1.8, iso500, photography, award-winning works --s 1000 --ar 3:4 --niji 5 --style expressive

古风中式美女，穿着浅绿色、白色和灰色搭配的裙子，头戴首饰和珠宝，发簪、手镯、项链、金、银镶玉制品，细腻的质地，色彩温暖，精致的脸，优雅的姿势，半身像，浅色背景，焦距 50，光圈 1.8，ISO500，获奖摄影作品 --s 1000 --ar 3:4 --niji 5 --style expressive

E211 - 李舒明

关键词：

portrait photography, clear and defined facial features, happy, excited, solid and clean color background, in combination with stunning epic Chinese ancient palace theme, adorned with gemstone hairpins, pearl hair clasps, golden crowns, earrings, jade pendants, bracelets, utilizing morandi color palette, post-processed with photoshop, and illustrated in the style of new art movement, 32k uhd --ar 3:4 --niji 5 --style expressive

人像摄影，清晰明确的五官，快乐，兴奋，纯净的彩色背景，结合令人惊叹的、史诗般的中国古代宫殿主题，装饰着宝石发夹、珍珠发扣、金冠、耳环、玉坠、手镯，利用莫兰迪调色板，用 Photoshop 进行后期处理，并以新艺术运动的风格进行说明，32K 超高清 --ar 3:4 --niji 5 --style expressive

M606－造字工人

关键词：

this is a portrait of an ancient Chinese noble beauty, a beautiful face, gorgeous Chinese ming dynasty costumes, take a lot of exquisite ancient Chinese jewelry ard accessories, rebecca guay style, light beige and light yellow, extreme rich detail, clamp, smile, serene, from side, left view, highly detailed --ar 9:16 --niji 5

这是一幅中国古代贵族美人的肖像，美丽的面容，华丽的明代服饰，有很多精美的中国古代珠宝配饰，丽贝卡·盖伊的风格，浅米色和浅黄色，极丰富的细节，夹子，微笑，宁静，从侧面，左视图，非常详细 --ar 9:16 --niji 5

E201 - 婷仔

关键词：

a beautiful whole body hanfu girl, jewelry, knolling layout --niji 5 --style expressive

一个美丽的全身汉服少女，首饰，直角排列的布局 --niji 5 --style expressive

小怪兽

T293 – 方梦竹

关键词：

white deer with four antlers, four antlers, white deer, wearing hanfu, blind box toy, blender, 3d, Octane rendering --niji 5 --style expressive

四角白鹿，四个鹿角，白鹿，穿汉服，盲盒玩具，Blender，3D，Octane 渲染 --niji 5 --style expressive

F233－豆非非

关键词：

a cute ip cartoon monster, Chinese lion dance costumes, nezha form, an anthropomorphic long-haired mountainmonster, kawaii,oriental art painting, national warriors wearing blue and red clothes, Chinese painting art, unique monster illustrations. scorpion lions, featured animals, grotesque images, movie lights, cream color scheme. clean white background, advanced color scheme. pixar style, Octane rendering,zbrush, super-realistic, high details --ar 2:3 --niji 5 --style expressive

一个可爱的卡通 IP 怪物，中国舞狮服装，哪吒形态，拟人化的长毛山怪，可爱，东方艺术绘画，穿着蓝色和红色衣服的民族勇士，中国绘画艺术，独树一帜的怪物插图。蝎子狮子，特色动物，怪诞的图像，电影灯光，奶油色方案。干净的白色背景，高级的配色方案。皮克斯风格，Octane 渲染，Zbrush，超逼真，高细节 --ar 2:3 --niji 5 --style expressive

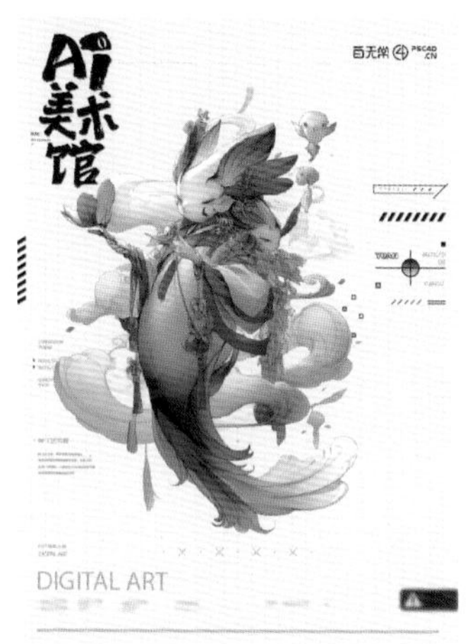

R856 -Acker 设计

关键词：

帝江：拟人化可爱的中国怪兽，形状像个布袋，整个身体红得像一团火，六只脚，两对翅膀，没有耳朵、鼻子和嘴巴，跳着舞蹈，古风舞蹈，敦煌画风，可爱的娃娃 IP，配色高级，色彩丰富，倪传婧插画风格，中国民间艺术，水墨渲染，黄金时代 / 新艺术时代，Octane 渲染，3D，细腻的细节，逼真的渲染，白色背景

精卫：拟人化中国飞鸟神兽，飞鸟，海燕，背后有一双翅膀，有着漂亮的羽毛，像人一样，拿着麦克风，尽情歌唱，敦煌画风，可爱的芭比娃娃，配色高级，色彩丰富，水墨渲染，Octane 渲染，3D，细腻的细节，逼真的渲染，白色背景

陆吾：拟人化的中国怪兽，人身，人脸，虎尾，背后有一双翅膀，外形像人，拿着贝斯，尽情舞动，动作夸张，乐队贝斯手，敦煌画风，金黄色的服装，可爱的娃娃 IP，配色高级，色彩丰富，半透明玻璃材质，倪传婧插画风格，水墨渲染，黄金时代 / 新艺术时代，Octane 渲染，3D，细腻的细节，逼真的渲染，白色背景

验证码：31236

E211－李舒明

关键词：

a weird-looking cute fat sheep, monster, surrounded by auspicious clouds, abstraction, in the style of classic of mountains and rivers pop mart-style, full body, wearing elaborate embroidered costumes, red, white and gold, hair, frayed, combined with the theme style of ancient Chinese mythology, red background, Octane rendering and blender rays of shimmering light, high quality, ultra-high definition, 32k uhd --ar 3:4 --niji 5 --style expressive

一只外形怪异可爱的肥羊，怪兽，祥云环绕，抽象，《山海经》风格和泡泡玛特风格，全身，穿着精致的刺绣服装，红色、白色和金色的头发，磨损，组合中国古代神话主题，红色背景，Octane 渲染并混合闪烁的光线，高质量，超高清，32K 超高清 --ar 3:4 --niji 5 --style expressive

T293－方梦竹

关键词：

white deer with four antlers, four antlers, white deer, blind box toy, blender, 3d, Octane rendering --niji 5 --style expressive

四角白鹿，四个鹿角，白鹿，盲盒玩具，Blender，3D，Octane 渲染 --niji 5 --style expressive

U931－侧柏

关键词：

a cool ip cartoon monster in dunhuang costume, movie card lighting, cream color scheme, clean white background, advanced color scheme, pixar style, surrealism, 3d, Octane rendering , more detail --ar 2:3

一个很酷的穿着敦煌服装的 IP 卡通怪物，电影卡片灯光，奶油色方案，干净的白色背景，高级配色方案，皮克斯风格，超现实主义，3D，Octane 渲染，更多细节 --ar 2:3

包装

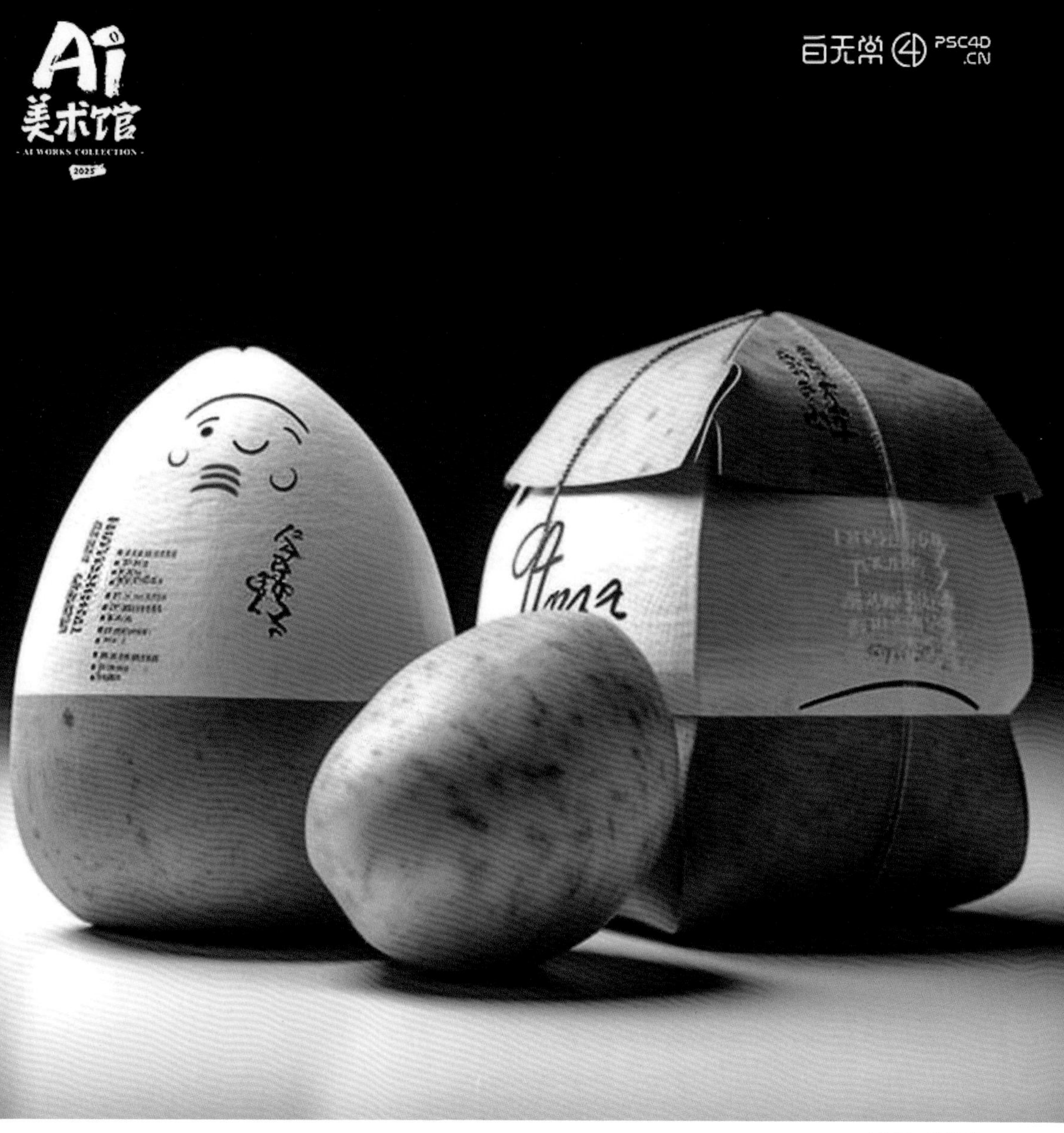

U907 - 鸭头君

关键词：

a potato package, creative, clean and bright background, cinematic light effects -- ar 3:2 --q 2 --s 750 --v 5

一个土豆包装，创意，干净明亮的背景，电影般的灯光效果 -- ar 3:2 --q 2 --s 750 --v 5

F259 – 郭亚丽

关键词：

moon cake gift box packaging design, huge miniature scene, clear glass texture, 3d paper cut art, embossed art, sense of hierarchy, flowing and breathable, complicated details, subtle tonal gradations, solid color background, epic light and shadow, i can't believe how beautiful this is, sony fe 12 – 24mm f/ 2. 8 gm, Octane rendering, high quality --ar 3:4 --q 2 --v 5.1 --style raw

月饼礼盒包装设计，巨大的微缩场景，清晰的玻璃质感，3D 剪纸艺术，浮雕艺术，层次感，流畅透气，复杂的细节，微妙的色调渐变，纯色背景，史诗般的光和影子，我简直不敢相信这是多么美丽，索尼 FE 12–24mm F/2.8GM，Octane 渲染，高质量 --ar 3:4 --q 2 --v 5.1 --style raw

Q783－罗洁

关键词：

bread packaging design, shaped packaging, rabbit shape, white, unique, high quality, white background --ar 3:4 --upbeta --q 2 --v 5

面包包装设计，定型包装，兔子形状，白色，独特，高品质，白色背景 --ar 3:4 --upbeta --q 2 --v 5

S871 - 依米

关键词：

inflatable, colored clear plastic, school bag, dieter rams, bauhaus, industrial design, pastel colors, good gloss, super close-up, white background, studio lighting --ar 3:4

充气，彩色透明塑料，书包，迪特 · 拉姆斯，包豪斯，工业设计，色彩柔和，光泽度好，超级特写，白色背景，工作室照明 --ar 3:4

Q783 - 罗洁

关键词：

candy packaging design, shaped packaging, rabbit shape, white, unique, high quality, white background --ar 3:4 --upbeta --q 2 --v 5

糖果包装设计，定型包装，兔子形状，白色的，独特的，高质量，白色背景 --ar 3:4 --upbeta --q 2 --v 5

V996- 叨叨

关键词：

meat packaging design, shaped bamboo packaging, simple theme, white background, angle, special effect, photo - realistic shadow

肉类包装设计，竹形包装，简单的主题，白色背景，角度，特效，照片般逼真的阴影

L573 - 王龙昊

关键词：

strawberry packaging design, special-shaped carton, interesting shape --q 2 --v 5

草莓包装设计，异形纸箱，趣味造型 --q 2 --v 5

N648－胡钱锋

关键词：

high transparency bag, white plastic outer packaging, gundam, designed by dieter rams, high detail, 4k, industrial design, white background, studio lighting, dieter rams, bauhaus --ar 3:4 --s 150 --q 2 --v 5.1

高透明袋子，白色塑料外包装，高达，由迪特·拉姆斯设计，高细节，4K，工业设计，白色背景，工作室照明，迪特·拉姆斯，包豪斯 --ar 3:4 --s 150 --q 2 --v 5.1

L551 - 时梦雅

关键词：

mushroom packaging design, boletus, special-shaped carton packaging, pentawards, interesting shape, cute --ar 3:4 --q 2 --niji 5

菌菇类包装设计，牛肝菌，异形纸盒包装，全球包装设计大奖赛，趣味造型，可爱 --ar 3:4 --q 2 --niji 5

三视图

A039－潮玩盲盒人物设计

关键词：

super cute girl, fluorescent translucent holographic jacket, blind box, pop mart design, full body, diamond luster, metallic texture, exaggerated expressions and movements, bright light, clay material, precision mech parts, close-up intensity, 3d, ultra-detailed, c4d, Octane rendering, blender, front side back three views, 8k, hd --ar 16:9 --niji 5 --style expressive

超级可爱的女孩，荧光半透明全息夹克，盲盒，泡泡玛特设计，全身，钻石光泽，金属质感，夸张的表情和动作，明亮的光线，黏土材质，精密的机械零件，强特写，3D，超多细节，C4D，Octane 渲染，Blender，正侧后三视图，8K，高清 --ar 16:9 --niji 5 --style expressive

A032 – 付丹丹

关键词：

front side back three views, back, front, side , blind box, a cute cartoon three views , protective goggles ::1, dark brown hat, orange top, protective goggles, full body shot, the front view, the side view, the back view, 16k, ip girl wearing a hiking suit, goggles, lapel top, China

正侧后三视图，背面，正面，侧面，盲盒，一个可爱的卡通三视图，护目镜 ::1，深棕色帽子，橙色上衣，护目镜，全身照，正视图，侧视图，后视图，16K，穿着登山服的女孩 IP，护目镜，翻领上衣，中国

G324－飞鸟 fia

关键词：

three views, full body shot, the front view, the side view, the back view, a cute cartoon ip minority girl, wearing a traditional ethnic minority clothing, wearing longboots,brown skin, flushed face, wearing complex decorations of agate turquoise, honey wax, coral, shells, light background, 3d, chibi, Octane rendering, bright movie light, advanced color scheme, high detail

三视图，全身拍摄，正视图，侧视图，后视图，一个可爱的卡通 IP 少数民族女孩，穿着少数民族传统服饰，穿着长筒靴，棕色皮肤，红润的脸，戴着复杂的玛瑙绿松石装饰，蜜蜡，珊瑚，贝壳，浅色背景，3D，赤壁，Octane 渲染，明亮的电影灯光，高级的配色方案，超多的细节

F236 – 何志昊

关键词：

three front and rear views, 3d toys, bubble gun girls shooting bubbles, fashion clothes, fashion shorts, fashion shoes, fashion posing, chibi, fluorescent translucent, luminous body, bauhaus, color, plastic, transparent, product design, complex mechanical parts, awe inspiring lighting, 3d, digital art, translucent plastic bubble gum, close-up, super detailed, c4d, Octane rendering, blender, high definition, full body, front view, rear view, side view, simple background --ar 16:9 --iw 2 --niji 5

前后三视图，3D 玩具，泡泡枪女孩射击泡泡，时尚服装，时尚短裤，时尚鞋，时尚姿势，赤壁，荧光半透明，发光体，包豪斯，色彩，塑料，透明，产品设计，复杂的机械部件，令人激动的灯光，3D，数字艺术，半透明塑料泡泡糖，特写，超强细节，C4D，Octane 渲染，Blender，高清，全身，前视图，后视图，侧视图，简单的背景 --ar 16:9 --iw 2 --niji 5

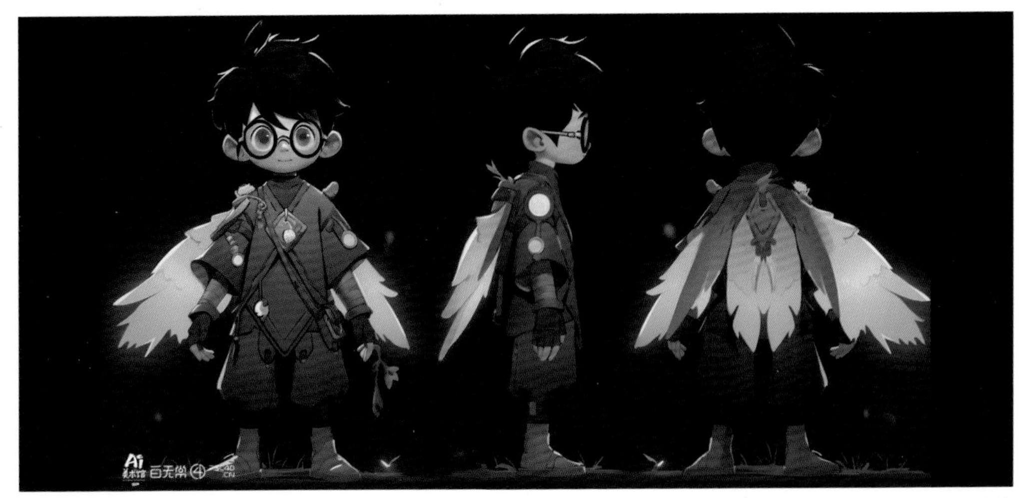

F233－豆非非

关键词：

cute Chinese boy with personality, with glasses, wave magic sticks, luminous body,color, plastic, transparent, product design, complex mechanical parts, wearing cute boots, full details, feathers, sunstone, forest background, soft movie lighting, disney pixar style characters, surrealism, 3d renderer, Octane rendering, ultra fine, ultra high definition, fashionable toys, complex details --ar 16:9 --q 2 --niji 5

可爱的有个性的中国男孩，戴着眼镜，挥动着魔棒，发光体，色彩，塑料，透明，产品设计，复杂的机械部件，穿着可爱的靴子，完整的细节，羽毛，日光石，森林背景，柔和的电影照明，迪士尼皮克斯风格的人物，超现实主义，3D渲染，Octane 渲染，超精细，超高清，时尚玩具，复杂细节 --ar 16:9 --q 2 --niji 5

B066－蔡旋

关键词：

reference art, side by side view, character art sheet, cute little girl character design in the style of ghibli, by disney, by pixar, flora --uplight --ar 16:9 --niji 5

参考艺术，并排视图，角色艺术表，吉卜力风格的可爱小女孩角色设计，迪士尼，皮克斯，花神弗洛拉 --uplight --ar 16:9 --niji 5

T285 - 文龙

关键词:

a cute personality dragon, front side back three views, walking on white auspicious clouds, wearing ming dynasty official clothes, pinterest, light background, auspicious, round, abnormal details, soft movie lighting, disney pixar style characters, surrealism, 3d renderer, Octane rendering, ultra fine, ultra high definition, fashionable toys, complex details --ar 16:9

一条个性可爱的龙，正侧后三视图，行走在白色祥云上，身披明代官服，Pinterest 网站，浅色背景，吉祥，圆形，异常的详细，柔和的电影灯光，迪士尼皮克斯风格人物，超现实主义，3D 渲染，Octane 渲染，超精细，超高清，时尚玩具，复杂细节 --ar 16:9

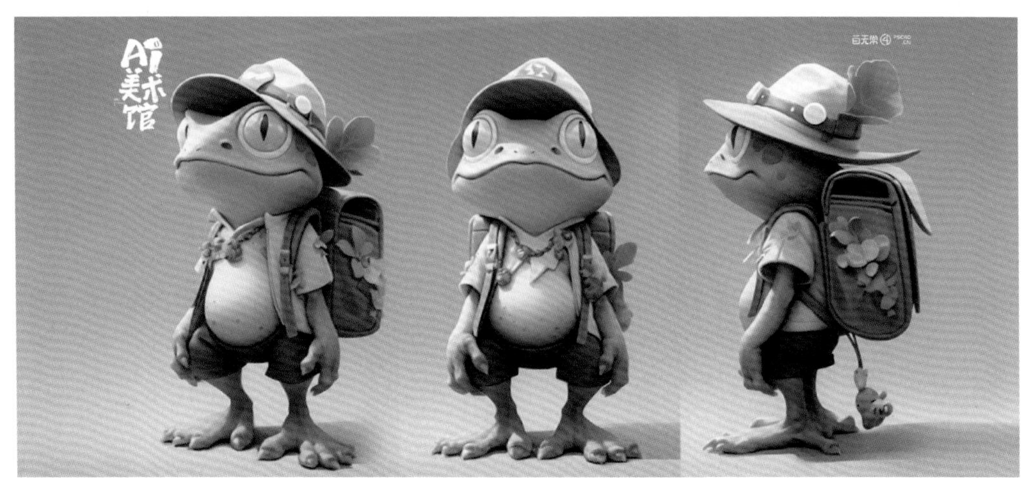

U912 - 吹啊吹

关键词:

a very beautiful frog, wearing a hat, blue shorts, trendy shoes, tourist frog, anthropomorphic design, fun and full of artist design, front side back three views, c4d, 3D vision, Octane rendering, zbrush, blender, 8K, HD --ar 16:9 --niji 5 --style expressive --s 180

一只非常漂亮的青蛙，戴着帽子，穿着蓝色短裤，时髦的鞋子，旅游青蛙，拟人化的设计，趣味十足艺术设计，正侧后三视图，C4D, 3D 视觉，Octane 渲染，ZBrush，Blender，8K，高清 --ar 16:9 --niji 5 --style expressive --s 180

U913 – ones

关键词：

a cute cartoon ip girl, wearing a dinosaur head gear, clean brown background, trendy shoes, white color scheme, pixar style, 3d, Octane rendering, bright cinematic light, three-quarter angle, advanced color scheme, high detail, front side back three views

一个可爱的卡通 IP 女孩，戴着恐龙头套，干净的棕色背景，时髦的鞋子，白色配色，皮克斯风格，3D，Octane 渲染，明亮的电影光线，四分之三角度，高级的配色方案，高细节，正侧后三视图

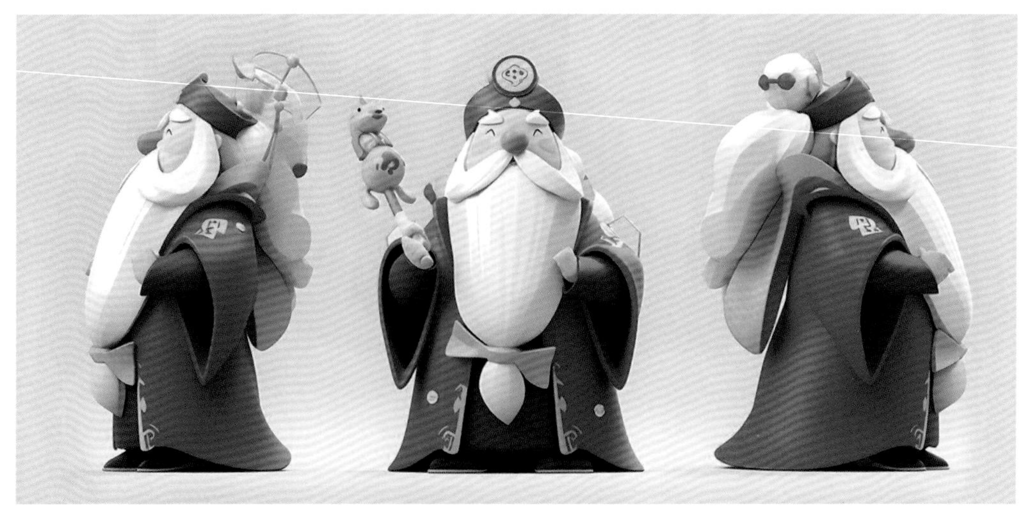

O690 – HuaJian

关键词：

front side back three views, god of wealth icon, super cute, behance, icon design, ui, bright, high – key lighting, white minimalist background, blind box, pop mart, 8k, best quality and super details, c4d, 3d --ar 16:9 --q 5 --niji 5 --style expressive

正侧后三视图，财神图标，超级可爱，Behance 网站，图标设计，UI，明亮，高级照明，白色极简主义背景，盲盒，泡泡玛特，8K，最佳质量和超多细节，C4D，3D --ar 16:9 --q 5 --niji 5 --style expressive

A027 – 于晓雯

关键词：

a cool little cartoon girl in a hoodie, hat, ip, curly hair, clean background, boots, cream color scheme, 3d, Octane rendering, bright movie lights, three-quarter view, advanced color scheme, high detail, front and back view --ar 16:9 --q 2 --niji 5

一个穿着连帽衫的炫酷卡通小女孩，帽子，IP，卷发，干净的背景，靴子，奶油配色方案，3D，Octane 渲染，明亮的电影灯光，四分之三视图，高级的配色方案，高细节，前后视图 --ar 16:9 --q 2 --niji 5

D 组 – 155 – 宁檬

关键词：

humanoid cat warrior, samurai sword, armor, three views of a cartoon image, generate three views, front view, wear straw shoes, the side view and the back view, wide shot --ar 16:9 --s 180 --niji 5 --style expressive

人形猫战士，武士刀，盔甲，卡通形象三视图，生成三视图，正视图，穿草鞋，侧视图和后视图，广角镜头 --ar 16:9 --s 180 --niji 5 --style expressive

C100－三视图・洛嘉仙子

关键词：

一个可爱的卡通 IP 天使，手办盲盒，IP 人物，白色长发，水晶王冠，水晶鞋，彩虹长裙，正侧后三视图，前视图，侧视图，背部身体形象 ::2，全身图像，浅色背景，美丽的脸，透明的挽歌衣服，CG，3D，C4D，Octane 渲染，8 头身比例，白色羽毛，薄纱乳胶，激光，彩虹色，全息图，王凌厚涂画风，金白银，柔焦，明亮的电影光线，明暗对比，8K

火把节

A015－贺小娇＋执着

关键词：

torch festival, large sacrificial sites, altar, ink painting, an expansive view, happy, high-definition, high-resolution, high details, clear faces, exquisite facial, 8k

火把节，大型祭祀场所，祭坛，水墨画，视野开阔，快乐，高清，高分辨率，高细节，面部清晰，五官精致，8K

I409 - 小六一

关键词：

around the bonfire, dancing, by andrew atroshenko, the dynamic composition showcases the joyful and smooth movements of the carnival dance, with vivid colors enhancing the dramatic movements --ar 3:4 --s 180 --niji 5 --style expressive

围着篝火，跳舞，来自安德鲁·阿托申科，充满活力的构图展现了欢快而流畅的狂欢节舞蹈，以生动的色彩加深了活动给人的深刻印象 --ar 3:4 --s 180 --niji 5 --style expressive

A011 – 吴斌

关键词：

carnival, Chinese young girl with a torch, singing and dancing, dance of flames, dancing with torches, laughter, celebration, bonfire, Chinese traditional festival, night, mountain village, rich details, clear texture, cute, unreal engine --ar 3:4 --niji 5 --style expressive

狂欢，中国少女手持火把，载歌载舞，火焰之舞，拿着火把跳舞，欢声笑语，庆典，篝火，中国传统节日，夜晚，山村，细节丰富，纹理清晰，可爱，虚幻引擎 --ar 3:4 --niji 5 --style expressive

A032 – 付丹丹

关键词：

torch festival, illustration, anime, hyper quality, clear facial features, holds torches, dance around the fire, a girl stands in the middle with a torch. it has profound cultural connotations and is known as the "oriental carnival". complex details, volume light, art --ar 3:4 --niji 5 --style expressive

火把节，插画，动漫，高品质，五官清晰，手持火把，围着火堆跳舞，一个女孩手拿着火把站在中间。它有着深厚的文化内涵，被称为“东方狂欢节”。复杂的细节，体积光，艺术 --ar 3:4 --niji 5 --style expressive

G323 - 徐阳

关键词:

girls with torches, singing and dancing, dance of flames, dancing with torches, laughter, celebration, bonfire --niji 5 --style expressive

女孩们手持火把，载歌载舞，火焰之舞，火把之舞，笑声，庆典，篝火 --niji 5 --style expressive

A031 – 兔子 + 廖蔓菊

关键词：

traditional torch festival style poster, girls in ethnic costumes dancing together, burning fire, bonfire, ancient festival, local culture, traditional culture with sharp colors, light and dark contrasts, exquisite details, motion blur, fast movement, hd 8k --ar 3:4 --q 2 --niji 5

传统火把节风格的海报，女孩穿着民族服装一起跳舞，燃烧的火焰，篝火，古老节日，地域文化，色彩鲜明的传统文化，明暗对比，细节精致，运动模糊，快速移动，高清 8K --ar 3:4 --q 2 --niji 5

A006－罗敏敏

关键词：

traditional ethnic costumes, young men and women dance around the flames, holding torches beautiful women and handsome men, fire, bottom up view, night, Octane rendering, 3d --ar 3:4 --style expressive --niji 5 --s 750

传统民族服饰，青年男女围着火焰跳舞，手持火把的美丽女子和英俊男子，火，仰视视角，夜晚，Octane 渲染，3D --ar 3:4 --style expressive --niji 5 --s 750

I406 - 大敢

关键词：

event photography, torch, beautiful 22-years-old asian girl, dancing with eight friends surrounding the girl, skin texture and perfect eyes, smiling, dancing and having a good time, holding a torch, wooden torch, using sony alpha 7iv, 85mm f/ 5.6, campfire party lights --ar 9:16 --s 250 --v 5.1

活动摄影，火炬，美丽的 22 岁亚裔女孩，8 个朋友围着女孩跳舞，有质感的皮肤和完美的眼睛，微笑，跳舞跳得很开心，拿着火把，木制火把，使用索尼 Alpha 7IV 相机，85mm F/ 5.6，篝火晚会灯光 --ar 9:16 --s 250 --v 5.1

A001 – 火把节 2

关键词：

torch festival, people were playing in the streets with torches, Chinese style building street --ar 2:3 --niji 5 --style expressive

火把节，人们拿着火把在街上玩耍，中式建筑街道 --ar 2:3 --niji 5 --style expressive

森林音乐会

B056－何颖珊

关键词：

cute boy with elf ears and transparent wings smiling and playing flute, liquid metal flower, white background, 3d rendering, ghibli style, dappled lighting, soft lighting, bright lighting tilt shift,isometric, 100mm lens, 8k --ar 3:4 --niji 5 --style expressive

长着精灵耳朵和透明翅膀的可爱男孩微笑着演奏长笛，液态金属花，白色背景，3D 渲染，吉卜力风格，斑驳的灯光，柔和的灯光，明亮的灯光倾斜移动，等距的，100mm 镜头，8K--ar 3:4 --niji 5 --style expressive

B057 – 汪云

关键词：

a cool lamb, with a fluffy afro is wearing a hip-hop style chain, a hawaiian short-sleeved shirt, white shorts and martin boots, while playing the saxophone. the background is a solid color, with studio lighting and soft tones, oc rendering

一只头发蓬松的酷羊羔，戴着嘻哈风格的项链，穿着夏威夷短袖衬衫，白色短裤和马丁靴，在演奏萨克斯。背景为纯色，工作室灯光和柔和的色调，Octane 渲染

C092－王木木

关键词：

plants vs. zombies, playing orchestral instruments, full body, pop mart,best quality, c4d, blender, 3d models, toys, vivid colors, high resolution, lots of details, pixar, candy colors, big shoes, fashion trends, art --ar 2:3 --niji 5 --style expressive

植物大战僵尸，演奏管弦乐乐器，全身，泡泡玛特，最好的质量，C4D，Blender，3D 模型，玩具，鲜艳的颜色，高分辨率，大量细节，皮克斯，糖果的颜色，大鞋，时尚趋势，艺术 --ar 2:3 --niji 5 --style expressive

G335 – Becky 半盏

关键词：

primitive tribe to gardenia fairy, 3d girl artwork, chibi,stand in the center of the big gardenia, play the accordion:: 10, brown red skin, blonde hair, gardenia braid headdress, fairy ears, wearing a skirt made of leaves, rich facial expression, exaggerated playing the accordion, long shot, epic light and shadow, clean background, natural lighting, 8k, best quality, ultra – detail, 3d, c4d, blender, oc rendering, ultra hd, 3d rendering --ar 3:4 --niji 5 --style expressive

原始部落栀子花仙子，3D少女艺术品，赤壁，站在大栀子花中央，拉手风琴 :: 10，棕红色皮肤，金发，栀子花花环头饰，仙女耳朵，穿着树叶做的裙子，丰富的面部表情，夸张的手风琴演奏，长镜头，史诗般的光影，干净的背景，自然光，8K，最佳质量，超强细节，3D，C4D，Blender，Octane 渲染，超高清，3D 渲染 --ar 3:4 --niji 5 --style expressive

K540 - Dilidili 敦煌音乐会

关键词：

the full body 3d artwork of cute little girl, dunhuang flying goddess, apsaras statue, dress with ribbons dancing, playing musical instruments, Chinese style, hanfu, classical, exquisite clothing, exquisite patterns, exquisite decorations, chibi, clay morphism, pastel color, c4d,oc rendering, ultra detailed, clean background --ar 2:3 --niji 5 --style expressive

可爱小女孩的全身 3D 艺术作品，敦煌飞天女神，飞天雕像，丝带舞服，演奏乐器，中式，汉服，古典，精致服饰，精美图案，精美装饰，赤壁，黏土材质，色彩柔和，C4D，Octane 渲染，超强细节，干净的背景 --ar 2:3 --niji 5 --style expressive

0719 – 木源海

关键词：

ip design, complete body, transparent holographic fruit composed in blind box style, an anthropomorphic grape holding a guitar playing the piano, holographic, fluorescent, transparent glass material, ethereal and dreamy style, playful and colorful depiction, clean background, oc rendering, blender, high quality, 32k, uhd --q 2 --niji 5 --style expressive

IP 设计，完整的身体，盲盒风格的透明全息水果，一个拟人化的葡萄拿着吉他弹着钢琴，全息的，荧光的，透明的玻璃材质，空灵梦幻的风格，俏皮多彩的描绘，干净的背景，Octane 渲染，Blender，高品质，32K，超高清 --q 2 --niji 5 --style expressive

Q777 - Matrix_·肆贰

关键词：

cute monster wearing Chinese martial arts costumes, Chinese wuxia. feminine, girl. water elemental spirit, Chinese painting style, cartoon ip, movie lights. clean white background, advanced color scheme, kung fu, Octane rendering, full body, surrealism, 3d trendy blind box toys, award winning

穿着中国武术服装的可爱怪物，中国武侠。女性的，女孩。水元素神韵，中国画风格，卡通 IP，电影灯光。干净的白色背景，高级配色，武术，Octane 渲染，全身，超现实主义，3D 潮流盲盒玩具，获奖

R851- 小朱

关键词：

a cute anthropomorphic cabbage band drummer, music vocalist, long white curly hair, stylish clothes, long legs and super shorts, stylish red martin boots, exaggerated drumming action, neon lights, ip, premium feel, silver and purple, c4d, oc rendering, particle effects, blind box, full body, weird images, solid color background, concept art, delicate details, silhouette --niji 5 --style expressive --s 400

一个可爱的拟人化卷心菜乐队鼓手，音乐歌手，长长的白色卷发，时尚的衣服，长腿和超短裤，时尚的红色马丁靴，夸张的击鼓动作，霓虹灯，IP，高级感，银色和紫色，C4D，Octane 渲染，粒子效果，盲盒，全身，怪异图像，纯色背景，概念艺术，精致细节，轮廓 --niji 5 --style expressive --s 400

R813－宋伟健

关键词：

Chinese style comics, Chinese classical costumes,playing the pipa, Chinese style, full body photo, light blue shoes, light blue hair, ultra-realistic portrait, ray tracking,handsome ancient Chinese girl, sunshine on his face, blue sky, big white clouds, magic engine 5 style. sculpture, transcendence, sabatier filter --ar 9:16 --niji5 --style expressive --s 400

中国风漫画，中国古典服饰，拨弄琵琶，中式，全身照，浅蓝色鞋子，浅蓝色的头发，超现实的肖像，光线的追踪，漂亮的古代中国女孩，阳光照在脸上，蓝色天空，大白云，虚拟引擎 5 风格。雕塑，超越，萨巴蒂尔过滤器 --ar 9:16 --niji5 --style expressive --s 400

M607 - Yuki

关键词：

a（水果）is wearing sneakers and holding a saxophone, surrealism, 3d, behance, pinterest, trendy blind box toys, 3d blind box, disney style, 8k, best quality --ar 3:4 --niji 5 --style expressive

一种(水果)穿着运动鞋和拿着萨克斯管乐器，超现实主义，3D，Behance网站，Pinterest网站，时尚的盲盒玩具，3D盲盒，迪士尼风格，8k，最高质量 --ar 3:4 --niji 5 --style expressive

D165 - 橘籽童學

关键词：

cute elf elements, playing instruments with a smile, full body, complex hair accessories, dunhuang flying sky, Chinese classical instruments, clean background, long lens, clay material, advanced color matching, 3d, 3d rendering, ip modeling, Octane rendering --ar 3:4 --niji 5 --style expressive --s 50

可爱的精灵元素，带着微笑演奏乐器，全身，复杂的发饰，敦煌飞天，中国古典乐器，干净的背景，长镜头，黏土材质，高级的配色方案，3D，3D 渲染、IP 建模、Octane 渲染 --ar 3:4 --niji 5 --style expressive --s 50

创意产品

T304 ~ Wang

关键词：

tall electric fan, sunflower, translucent melt, inflatable, bauhaus style, high detail, 8k hd, industrial design, detailed, clean white background, advanced color matching, studio lighting --ar 9:16

高大的电风扇，向日葵，半透明融化，可充气，包豪斯风格，高细节，8K 高清，工业设计，细节，干净的白色背景，高级的配色方案，工作室照明 --ar 9:16

D177 – Suiseiesk

关键词：

design a chandelier with sunflower look, bionic::2 design, transparent material, studio lighting, behance, blender, Octane rendering, high detail, 8k, detail light color --ar 16:12 --s 750 - image #4

设计一个向日葵外观的枝形吊灯，仿生 ::2 设计，透明材料，工作室照明，Behance 网站，Blender，Octane 渲染，高细节，8K，细致处理光色 --ar 16:12 --s 750 - image #4

F254 – Astart

关键词：

machine dog and horn sound combination, soluble transparent shell, tripoli, Spain style, new york subway city background, clear internal structure, 8k, product photography, premium feeling. art, unreal style, ue5 --q 2 --s 250 --v 5.1 – image #4

机器狗和喇叭声音组合，可溶透明外壳，的黎波里，西班牙风格，纽约地铁城市背景，清晰的内部结构，8K，产品摄影，高级感。艺术，虚幻风格，虚幻引擎 5 --q 2 --s 250 --v 5.1 – image #4

C119 – 既死又活的 Win

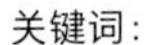

关键词：

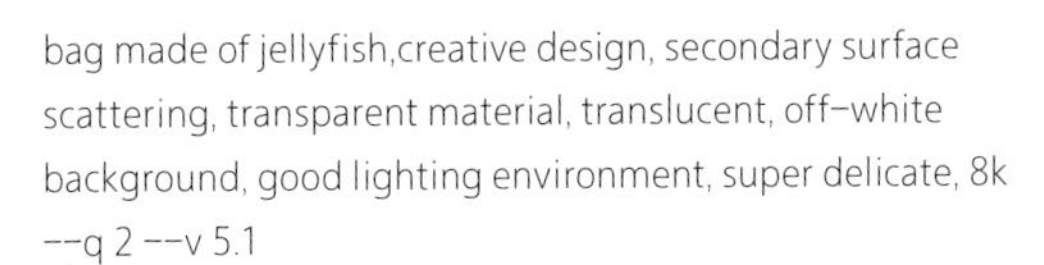

bag made of jellyfish,creative design, secondary surface scattering, transparent material, translucent, off-white background, good lighting environment, super delicate, 8k --q 2 --v 5.1

由水母制成的包，创意设计，次表面散射，透明材料，半透明，米白色背景，良好照明环境，超精致，8K --q 2 --v 5.1

R828 – hm

关键词：

sofa, hairy, hamburger shape, cute, disney style, white background, 3d, Octane rendering --q 2 --v 5

沙发，毛茸茸的，汉堡包形状，可爱，迪士尼风格，白色的背景，3D，Octane 渲染 --q 2 --v 5

C119 – 既死又活的 Win

关键词：

bus shape, made of wool felt, wool material, bright and clean lighting environment, beige small and fresh, ultra delicate, miniature shift axis photography --q 2--v 5.1

公交车形状，由羊毛毡制成，羊毛材料，明亮干净的照明环境，米色小清新，超细致，微型移轴摄影技术 --q 2--v 5.1

E195－清风

关键词：

shelf drum, neon color, translucent melt, by dieter rams, dieter rams designer, bauhaus style, high detail, 8k hd, industrial design, detailed, clean white background, advanced color matching, studio light --ar 3:4 --v 5.1 --s 400

架子鼓，霓虹色，半透明融化，由迪特·拉姆斯设计，设计师迪特·拉姆斯，包豪斯风格，高细节，8K 高清，工业设计，细节，干净的白色背景，高级的配色方案，工作室灯光 --ar 3:4 --v 5.1 --s 400

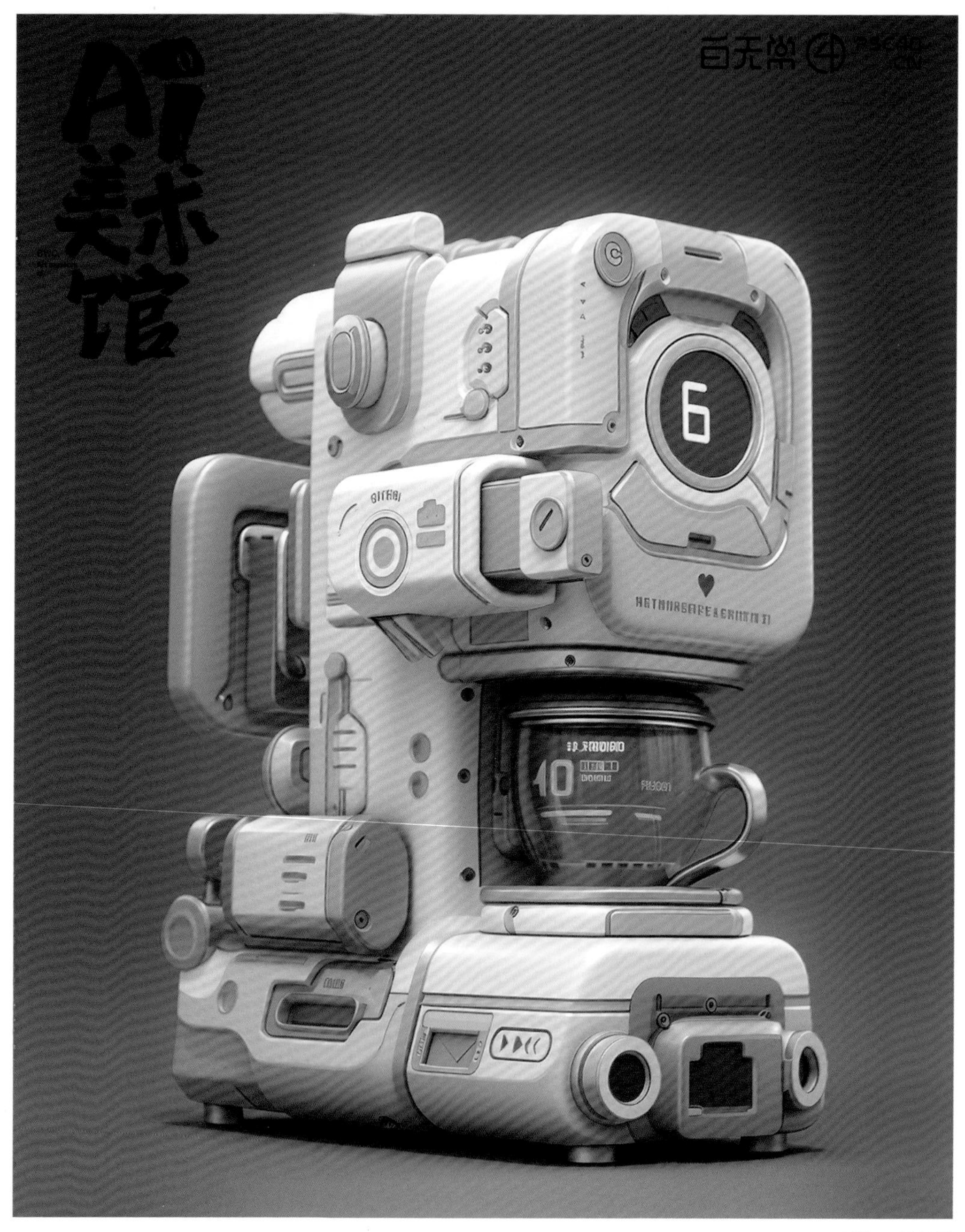

L551－时梦雅

关键词：

concept coffee machine, futuristic machinery, surrealism, light effect art laser printed stamped camellia pattern, minimalism, bauhaus style, high detail, 8k hd, industrial design, detailed, clean background, advanced color matching, photographic lighting --q 2 --v 5 --niji 5

概念艺术咖啡机，未来机械，超现实主义，光效艺术激光器打印山茶花图案，极简主义，包豪斯风格，高细节，8K 高清，工业设计，细节，干净的背景，高级的配色方案，摄影照明 --q 2 --v 5 --niji 5

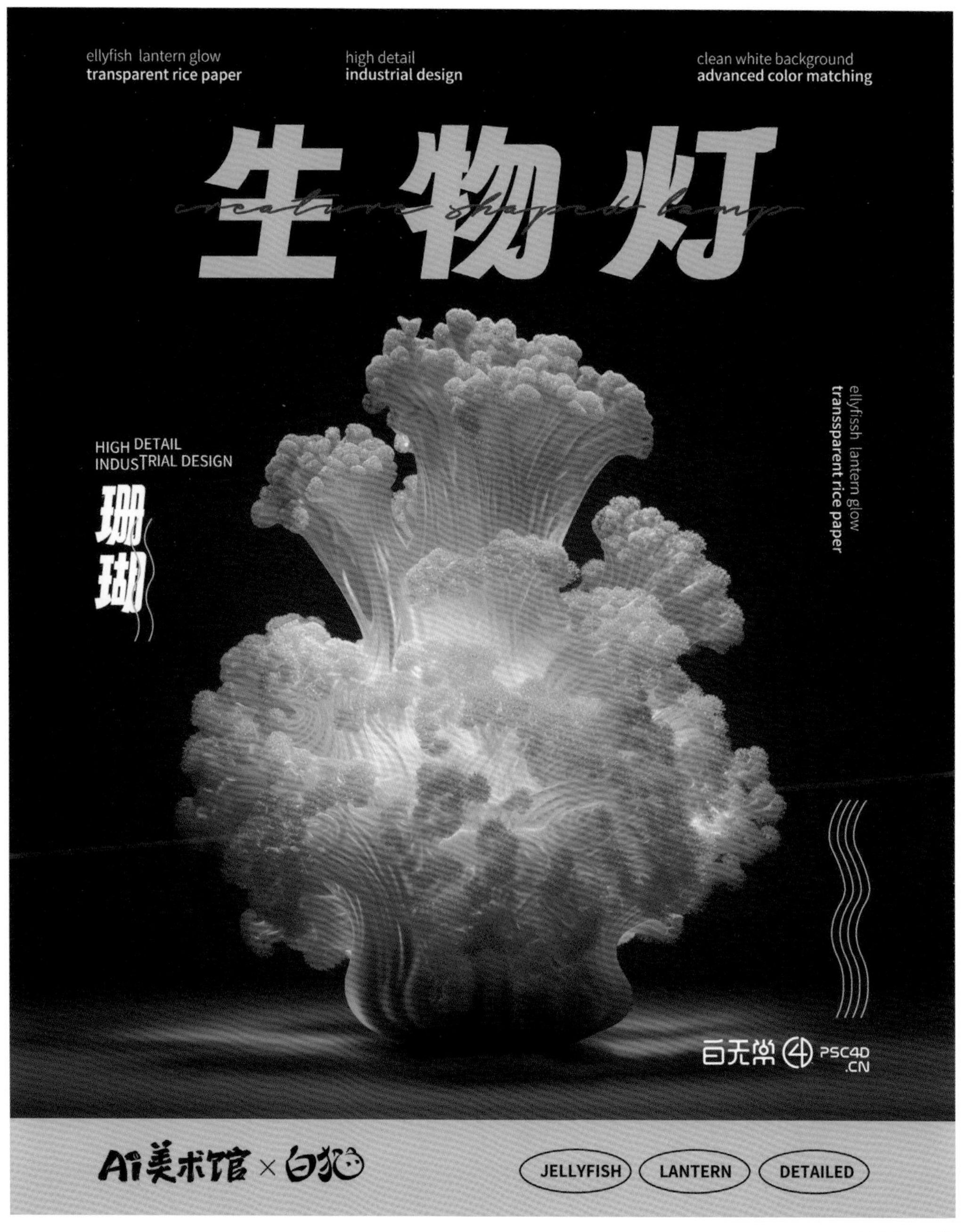

M617－白猫

关键词：

coral, lights, glow, translucent, high detail, 8k hd, industrial design, detailed, clean white background, advanced color matching, studio light --v 5.1 --q 2

珊瑚，灯光，发光，半透明，高细节，8K 高清，工业设计，细节，干净的白色背景，高级的配色方案，工作室灯光 --v 5.1 --q 2

T275 – sophia

关键词：

translucent glass molten body, laser effect, caustics, designed by dieter rams, industrial design, high detail, glowing, oc rendering --q 2 --v 5 – image #2

半透明玻璃熔融体，激光效果，焦散，由迪特 · 拉姆斯设计，工业设计，高细节，发光，Octane 渲染 --q 2 --v 5 – image #2

R843 - 哈尼叶子

关键词：

plastic, translucent molten body, high detail, Octane render, athletic shoes out of spherical objects of varying sizes, shoes in huge ball, 4k, 8k --ar 2:3 --v 5.1

塑料，半透明熔融体，高细节，Octane 渲染，由不同尺寸的球形物体制成的运动鞋，鞋在大球上，4K，8K --ar 2:3 --v 5.1

D170 - 皖豫

关键词：

rice cooker modeling, design, creativity, pentawards, Chinese ink painting, lacquer painting and oil painting printing, commercial photography, ancient architectural modeling, high-resolution photography, cool colors, canon camera iso400, 8k, oc rendering, high-resolution, high, high detail, award winning design, ultra-realistic, 3d rendering --ar 4:3 --niji 5 --style expressive

电饭煲造型，设计，创意，全球包装设计大奖赛，中国水墨画，漆画和油画印刷，商业摄影，古建筑建模，高分辨率摄影，冷色调，佳能相机 ISO400，8K，Octane 渲染，高分辨率，高细节，获奖设计，超现实，3D 渲染 --ar 4:3 --niji 5 --style expressive

R816- 罐头鱼

关键词：

low saturation perfume bottle, coloured plastic, Egyptian perfume, Egyptian style, Egyptian design elements, industria design, bauhaus, fine lustre, rich colours, plastic transparent, white background, studio lighting, 8k, clear glass

低饱和度的香水瓶，彩色塑料，埃及香水，埃及风格，埃及设计元素，工业设计，包豪斯，光泽细腻，色彩丰富，塑料透明，白色背景，工作室照明，8K，透明玻璃

V964 - 陈晖

关键词：

sofa, the front view, elegant black feather sofa, long black feather robe with thorns, yohji yamamoto style, white background, 3d, Octane rendering --q 2 --v 5

沙发，正视图，典雅的黑色羽毛沙发，带刺的黑色羽毛长袍，山本耀司风格，白色背景，3D，Octane 渲染 --q 2 --v 5

M592 - 生詸果

关键词：

furniture design, sofa, inflatable, transparent, in the forest, high detail, 8k hd, industrial design, detailed, advanced color matching, studio light --ar 2:3

家具设计，沙发，充气，透明，在森林里，高细节，8K 高清，工业设计，细节，高级的配色方案，工作室灯光 --ar 2:3

B062 – Pink 河马

关键词：

hippocampus made of transparent plastic paper, transparent color (pink) plastic paper, lantern art, lantern structure, structure with sharp edges and corners, structural – support, translucent, white background, photographic effect, realistic, simple, minimalist style, pure white empty background, top view, 16k --ar 3:4 --s 200 --q 2 --v 5

由半透明塑料制成的海马，半透明彩色（粉色）塑料纸，花灯艺术，灯笼结构，结构棱角分明，结构支撑，半透明，白色背景，摄影效果，写实，简单，极简风格，纯白色空背景，俯视图，16K --ar 3:4 --s 200 --q 2 --v 5

J456－冉

关键词：

transparent technology motorcycle, gradient translucent glass molten body, laser effect, caustics, designed by dieter rams, clear glass, minimalistic, high detail, glowing, white background, industrial design, studio lighting, c4d,oc renderer, clean shadows, 8k --q 2 --v 5

半透明技术摩托车，渐变半透明玻璃熔融体，激光效果，焦散，由迪特·拉姆斯设计，透明玻璃，极简主义，高细节，发光，白色背景，工业设计，工作室照明，C4D，Octane 渲染器，干净阴影，8K --q 2 --v 5

插画风格设计师

李伊宁

关键词：

stylish designers, flat design illustration, orange - white color, geometric shapes, graphic illustration, minimalism, clean fresh design, advanced color matching, flat style, minimalist art, abstract memphis, pinterest, dribbble, super detail --ar 2:3

时尚设计师，平面设计插画，橙白配色，几何造型，图形插画，极简主义，清新的设计，高级的配色方案，平面风格，极简艺术，抽象孟菲斯，Pinterest 网站，Dribbble 网站，超级细节 --ar 2:3

水果 + 产品组合

1. 水果与包

李伊宁

关键词：

a classy and elegant handbag, for the cover of vogue Italia, with some cut fruit scattered around, botanical decoration, top light, clean natural background, high saturation color scheme, bright product light, dappled shadows, studio lighting, sense of contrast, high precision, fine gloss, photography, hd, 8K, realism --ar 2:3

一个高档优雅的手提包，用于意大利杂志《VOGUE》的封面，四周散落着一些切好的水果，植物装饰，顶光，干净的自然背景，高饱和度的配色方案，明亮的产品光，斑驳的阴影，工作室灯光，对比感，高精度，精细光泽，摄影，高清，8K，现实主义 --ar 2:3

2. 水果与手机壳

李伊宁

关键词：

a classy and elegant phone, for the cover of vogue Italia, with some cut fruit scattered around, botanical decoration, top light, clean natural background, high saturation color scheme, bright product light, dappled shadows, studio lighting, sense of contrast, high precision, fine gloss, photography, hd, 8k, realism --ar 2:3

一部优雅时尚的手机，用于意大利杂志《VOGUE》的封面，周围散落着一些切好的水果，植物装饰，顶光，干净自然的背景，高饱和度的配色方案，明亮的产品光，斑驳的阴影，工作室照明，对比感，高精度，精细光泽，摄影，高清，8K，现实主义 --ar 2:3

3. 水果与口红

李伊宁

关键词：

a classy and elegant lipsticks, for the cover of vogue Italia, with some cut fruit scattered around, botanical decoration, top light, clean natural background, high saturation color scheme, bright product light, dappled shadows, studio lighting, sense of contrast, high precision, fine gloss, photography, hd, 8k, realism --ar 2:3

一款高雅精致的口红，用于意大利杂志《VOGUE》的封面，四周散落着一些切好的水果，植物装饰，顶光，干净自然的背景，高饱和度的配色方案，明亮的产品光，斑驳的阴影，工作室照明，对比感，高精度，精细光泽，摄影，高清，8K，现实主义 --ar 2:3

数字植物

李伊宁

关键词：

a futuristic beautifully organic flower, elegantly grown digital plant, hyper-artistic expression, film coating, metallic glass gloss, surrealist style, cool tones, light green jade and light purple colors, subtle glow, light background, studio lighting, macro nature photography, elegant wild lily, super detail, 8k --ar 9:16

一枝未来感十足的美丽有机花朵，优雅生长的数字植物，超艺术的表达，薄膜涂层，金属玻璃光泽，超现实主义风格，冷色调，浅绿色玉石和浅紫的颜色，微妙的光芒，明亮的背景，工作室照明，微距自然摄影，优雅的野百合，超级细节，8K --ar 9:16

昆虫

李伊宁

关键词：

a special beautiful insect, accurate insect specimens made of flower petals trash, nature inspired camouflage, detailed foliage, light orange and light green, goro fujita, ambrosius bosschaert,film lighting,zbrush,3d,redshift, hyperrealism,light background --ar 2:3 --s 250

一只特别美丽的昆虫，用花瓣废料制成的精确的昆虫标本，以自然为灵感的伪装，细致的叶子，浅橙色和浅绿色，藤田五郎，安布罗修斯 · 博斯查尔特，电影照明，ZBrush，3D，红移渲染器，超现实主义，浅色背景 --ar 2:3 --s 250

人像海报

李伊宁

关键词：

a beautiful digital man, with calm and elegant mood, trapped mood depiction, neo-brutalist style, ethereal details, album art, realistic skin texture, 8k

一个美丽的数字人，平静而优雅的情绪，被困的情绪描绘，新野兽派风格，空灵的细节，专辑艺术，逼真的皮肤纹理，8K

太空站

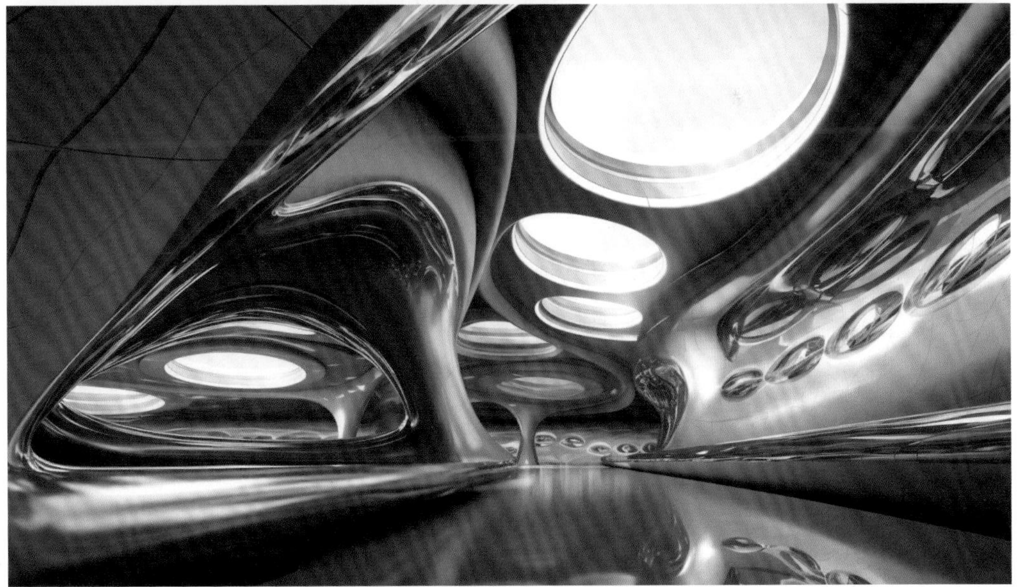

李伊宁

关键词：

futuristic space station appearance::1, futuristic architecture, light silver and dark, bold yet graceful, golden spiral composition, bruno taut, zaha hadid, extreme wide shot, overhead shot or bird's eye view, cold and warm contrast --ar 16:9

未来空间站外观 ::1，未来主义的建筑，浅银色和深色，大胆而优雅，金色螺旋形构图，布鲁诺 · 陶特，扎哈 · 哈迪德，极端广角镜头，俯视图或鸟瞰图，冷暖对比 --ar 16:9

仙人掌怪

李伊宁

关键词：

the cactus monster in the game, a cactus warrior holding the original weapon attack, detailed plant textures, first-person perspective attack shots, close-up shots, breathtaking moments shot capture, harry potter, pixar, 3d, blender, movie lighting, realistic style, realistic detail rendering, 8k --ar 2:3

游戏中的仙人掌怪物，仙人掌战士手持原始武器进行攻击，植物纹理细致，第一人称视角攻击镜头，特写镜头，惊险瞬间镜头捕捉，哈利波特，皮克斯，3D，Blender，电影照明，写实风格，逼真渲染细节，8K--ar 2:3

机甲小恐龙

李伊宁

关键词：

a robot of an little and cute android warrior dragon, attractive and beautiful eyes, in the style of rtx, mecha, mechanical wings, future creatures, proud and graceful, blink fire core, rich detailing, molecular, neon,gold light and rich light color scheme, in the style of cyberpunk, nature-inspired forms, 3d, blender, redshift --ar 7:10 --niji 5 --style expressive --s 180

一只小而可爱的机器战斗龙，迷人而美丽的眼睛，RTX 风格，光滑的机甲，机械，未来的生物，骄傲而优雅，闪烁的火芯，丰富的细节，分子，霓虹灯，金色光和丰富的灯光配色方案，赛博朋克风格，自然灵感的形式，3D，Blender，红移渲染器 --ar 7:10 --niji 5 --style expressive --s 180

室内环境

1. 儿童游戏室

李伊宁

关键词：

children's playroom hd photos, interior design, art, close-up, celluloid, creme wind, top view, bates masi architects, high quality photos, ultra hd, 8k --ar 3:2

儿童游戏室高清照片，室内设计，艺术感，特写，赛璐珞，奶油风，俯视视角，贝茨玛斯建筑和设计事务所，高质量照片，超高清，8K --ar 3:2

李伊宁

关键词：

high quality home interior environment living room photos local close-up, fresh log style, interior design, artistic sense, minimalism, overlooking perspective, oblique composition, john pawson, high quality photos, ultra hd, 8k --ar 3:2

高品质家居室内客厅照片局部特写，清新原木风格，室内设计，艺术感，极简主义，俯视视角，倾斜构图，约翰·帕森，高质量照片，超高清，8K --ar 3:2

以小见大

1. 仙人掌

李伊宁

关键词：

微缩景观，广告设计图像，以一些巨大的仙人掌雕塑为背景，以小见大的修饰风格，风格优美，细节丰富，氛围感强，传统文化主题风格，电影海报，梦幻场景，超现实的 3D 景观风格，电影照明

2. 玉米

李伊宁

关键词：

微缩景观，广告设计图像，以一些巨大的玉米雕塑为背景，以小见大的修饰风格，风格优美，细节丰富，氛围感强，传统文化主题风格，电影海报，梦幻场景，超现实的 3D 景观风格，电影照明

3. 多肉

李伊宁

关键词：

微缩景观，广告设计图像，以一些巨大的多肉雕塑为背景，以小见大的修饰风格，风格优美，细节丰富，氛围感强，传统文化主题风格，电影海报，梦幻场景，超现实的 3D 景观风格，电影照明

4. 面包

李伊宁

关键词：

微缩景观，广告设计图像，以一些巨大的面包雕塑为背景，以小见大的修饰风格，风格优美，细节丰富，氛围感强，传统文化主题风格，电影海报，梦幻场景，超现实的 3D 景观风格，电影照明

沙漠

李伊宁

关键词：

huge pale pink desert, ethereal landscape, soft light colors, hyper-realistic style, depth of field, extreme elevation, golden spiral composition, cinematic lighting, hd, 8k --ar 16:9

巨大的浅粉色沙漠，缥缈的风景，柔和的色彩，超写实风格，景深，极端的海拔，黄金螺旋构图，电影级的照明，高清，8K --ar 16:9

绘本

1. 如果我们不曾相遇

那一年
我17岁 它5岁
我们在生命里
相遇

那一年
我19岁 它7岁
我们是再也无法分割的
双翼
百无常 PSC4D.CN AI美术馆

那一年
我22岁 它10岁
我们并肩看过人间
无数风景

那一年
我26岁 它12岁
那一天
它始终没有出现

那一年
我32岁 它12岁
我带着不再变老的它
旅行
百无禁忌 PSC4D.CN AI美术馆

C124－K.k

关键词：

tight shot, the yellow forest, middle empty, jon klassen style, high details, dark colour, high contrast --ar 6:4 --niji 5 --style expressive

抓拍，黄色的森林，中间是空的，乔恩 · 克拉森风格，高细节，深色，高对比度 -- ar 6:4 --niji 5 --style expressive

a lovely old lady named ee, unhappy, in the forest, jon klassen style, playful, high details, dark colour, high contrast --ar 6:4 --iw 2 --niji 5 --style expressive

一个叫 EE 的可爱的老太太，不开心，在森林里，乔恩 · 克拉森风格，俏皮，高细节，深色，高对比度 --ar 6:4 --iw 2 --niji 5 --style expressive

a cute girl named ee hiding behind the tree, jon klassen style, playful, high details, dark colour, high contrast --ar 6:4 --iw 2 --niji 5 --style expressive

一个叫 EE 的躲在树后的可爱女孩，乔恩 · 克拉森风格，俏皮，高细节，深色，高对比度 --ar 6:4 --iw 2 --niji 5 --style expressive

a cute monster named momo, afraid, hiding behind a tree, jon klassen style, playful, high details, dark colour, high contrast --ar 6:4 --niji 5 --style expressive

一个叫 MoMo 的可爱怪物，害怕，躲在树后，乔恩 · 克拉森风格，俏皮，高细节，深色，高对比度 --ar 6:4 --niji 5 --style expressive

a cute girl named ee, a monster named momo, stand together, jon klassen style, playful, high details, dark colour, high contrast --ar 6:4 --niji 5 --style expressive

一个叫 EE 的可爱女孩，一个叫 MoMo 的怪物，站在一起，乔恩 · 克拉森风格，俏皮，高细节，深色，高对比度 --ar 6:4 --niji 5 --style expressive

a cute girl named ee, a monster named momo, dancing together, jon klassen style, playful,high details, dark colour, high contrast --ar 6:4 --iw 2 --niji 5 --style expressive

一个叫 EE 的可爱女孩，一个叫 MoMo 的怪物，一起跳舞，乔恩 · 克拉森风格，俏皮，高细节，深色，高对比度 --ar 6:4 --iw 2 --niji 5 --style expressive

a cute girl named ee, a monster named momo, sleeping together in the forest, jon klassen style, playful, high details, dark colour, high contrast --ar 6:4 --iw 2 --niji 5 --style expressive

一个叫 EE 的可爱女孩，一个叫 MoMo 的怪物，在森林里一起睡觉，乔恩 · 克拉森风格，俏皮，高细节，深色，高对比度 --ar 6:4 --iw 2 --niji 5 --style expressive

a cute girl named ee, a monster named momo, stand together on the top of the mountain, jon klassen style, playful, high details, dark colour, high contrast --ar 6:4 --iw 2 --niji 5 --style expressive

一个叫 EE 的可爱女孩，一个叫 MoMo 的怪物，一起站在山顶，乔恩 · 克拉森风格，俏皮，高细节，深色，高对比度 --ar 6:4 --iw 2 --niji 5 --style expressive

a cute girl named ee, in the daytime, in the forest, jon klassen style, playful, high details, dark colour, high contrast --ar 6:4 --iw 2 --niji 5 --style expressive

一个叫 EE 的可爱女孩，白天，在森林里，乔恩 · 克拉森风格，俏皮，高细节，深色，高对比度 --ar 6:4 --iw 2 --niji 5 --style expressive a cute girl named ee, unhappy, in the forest, jon klassen style, playful, high details, dark colour, high contrast --ar 6:4 --iw 2 --niji 5 --style expressive

一个叫 EE 的可爱女孩，不开心，在森林里，乔恩 · 克拉森风格，俏皮，高细节，深色，高对比度 --ar 6:4 --iw 2 --niji 5 --style expressive

a cute girl named ee, by the sea, jon klassen style, playful, high details, dark colour, high contrast --ar 6:4 --niji 5 --style expressive

一个叫 EE 的可爱女孩，在海边，乔恩 · 克拉森风格，俏皮，高细节，深色，高对比度 -- ar 6:4 --niji 5 --style expressive

a cute girl named ee, by the sea, jon klassen style, playful, high details, dark colour, high contrast --ar 6:4 --niji 5 --style expressive

一个叫 EE 的可爱女孩，在海边，乔恩 · 克拉森风格，俏皮，高细节，深色，高对比度 --ar 6:4 --niji 5 --style expressive

a lovely old lady named ee, unhappy, in the forest, jon klassen style, playful, high details, dark colour, high contrast --ar 6:4 --iw 2 --niji 5 --style expressive

一个名叫 EE 的可爱的老太太，不开心，在森林里，乔恩 · 克拉森风格，俏皮，高细节，深色，高对比度 --ar 6:4 --iw 2 --niji 5 --style expressive

a lovely old lady named ee, standing on the top of the mountain, jon klassen style, playful, high details, dark colour, high contrast --ar 6:4 --iw 2 --niji 5 --style expressive

一个名叫 EE 的可爱的老太太，站在山顶，乔恩 · 克拉森风格，俏皮，高细节，深色，高对比度 --ar 6:4 --iw 2 --niji 5 --style expressive

2. 造纸术

AI美术馆
1. 收集树皮、竹子、稻草
和废纸等富含纤维的原料。
百无尚 PSC4D .CN

AI美术馆
2. 剥离树皮、切碎竹子，整理稻草。
百无尚 PSC4D .CN

AI美术馆
3. 将原料浸泡于水中，
使纤维更易分离。
百无尚 PSC4D .CN

AI美术馆
4. 研磨浸泡过的原料，制作纸浆。
百无尚 PSC4D .CN

5. 使用造纸模具和筛网铺设纸浆。

6. 压潮湿的纸张，去除多余水分。

T280 – PPT

关键词：

the cover of a picture storybook: outside wooden house,a little white rabbit dressed in ancient coarse cloth clothing is reading a book --ar 3:2

一本图画故事书的封面：在屋子外，一只穿着古代粗布衣服的小白兔正在看书 --ar 3:2

page of a picture storybook: in the forest,a little white rabbit dressed in ancient coarse cloth clothing is gathering bark and bamboo,sorting straw and waste paper --ar 3:2

一本图画故事书的一页：在树林里，一只穿着古代粗布衣服的小白兔正在收集树皮和竹子，稻草和废纸 --ar 3:2

page of a picture storybook: in the forest,a little white rabbit dressed in ancient coarse cloth clothing is peeling bark,chopping bamboo and sorting straw --ar 3:2

一本图画故事书的一页：在树林里，一只穿着古代粗布衣服的小白兔正在剥离树皮，切碎竹子和整理稻草 --ar 3:2

page of a picture storybook: in the forest, a little white rabbit dressed in ancient coarse cloth clothing is pouring water into a large wooden barrel --ar 3:2

一本图画故事书的一页：在树林里，一只穿着古代粗布衣服的小白兔正在往一个大木桶里倒水 --ar 3:2

page of a picture storybook: inside a wooden house, a little white rabbit dressed in ancient coarse cloth clothing is scooping pulp onto a sieve --ar 3:2

一本图画故事书的一页：在一座木屋里，一只穿着古代粗布衣服的小白兔正在把纸浆舀到筛子上 --ar 3:2

page of a picture storybook: close-up shot, a basin and sieve are placed on a wooden table, with the outline of the paper faintly visible --ar 3:2

一本图画故事书的一页：特写镜头，一个盆和筛子放在木桌上，纸的轮廓隐约可见 --ar 3:2

page of a picture storybook: inside a wooden house, a little white rabbit dressed in ancient coarse cloth clothing is pressing a stack of paper with a wooden board, with water seeping out from the edges of the paper --ar 3:2

一本图画故事书的一页：在一个木屋里，一只穿着古代粗布衣服的小白兔正用木板压着一堆纸，水从纸边渗出来 --ar 3:2

page of a picture storybook: in the village, a little white rabbit dressed in ancient coarse cloth clothing is hanging paper on a drying rack to dry, sunlight shines on the paper --ar 3:2

一本图画故事书的一页：在村里，一只穿着古代粗布衣服的小白兔正把纸挂在晒衣架上晾干，阳光照在纸上 --ar 3:2

3. 木兰辞

女亦无所思

女亦无所忆

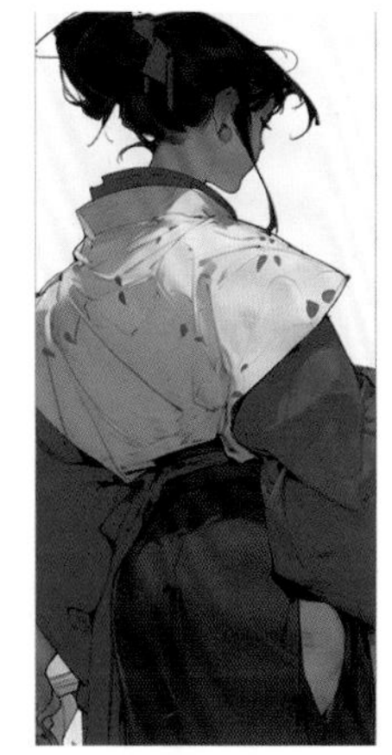

昨夜见军帖

可汗大点兵

军书十二卷
卷卷有爷名

阿爷无大儿
木兰无长兄

愿为市鞍马
从此替爷征

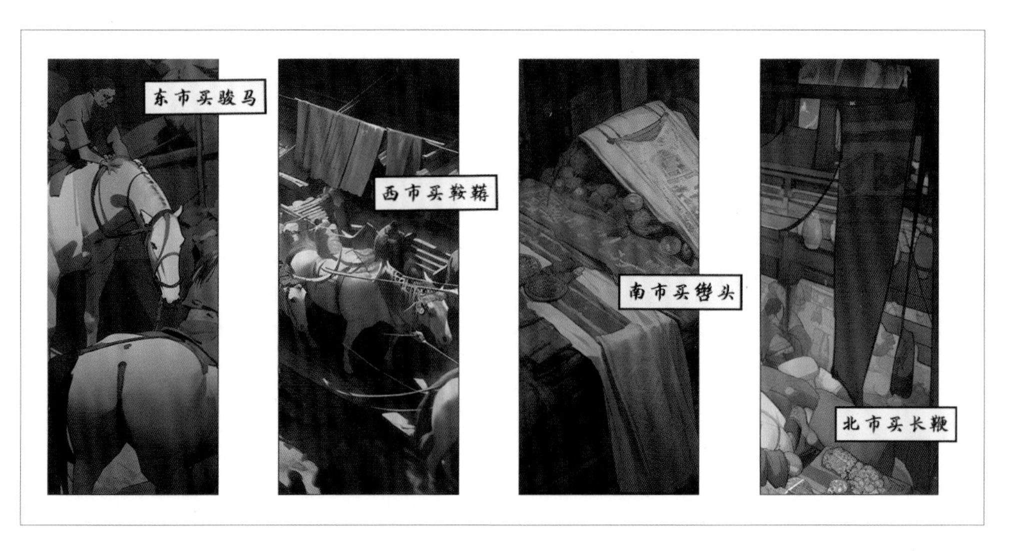
东市买骏马
西市买鞍鞯
南市买辔头
北市买长鞭

旦辞爷娘去

暮宿黄河边

不闻爷娘唤女声
但闻黄河流水鸣溅溅

旦辞黄河去
暮至黑山头

不闻爷娘唤女声
但闻燕山胡骑鸣啾啾

万里赴戎机

关山度若飞

朔气传金柝

寒光照铁衣

将军—

—百战死

壮士——

——十年归

归来见天子

天子坐明堂

花木兰
最高一等功
钦此
策勋十二转

赏赐百千强

可汗问所欲

木兰不用尚书郎

愿驰千里足

送儿还故乡

爷娘闻女来
出郭相扶将

阿姊闻妹来
当户理红妆

小弟闻姊来
磨刀霍霍向猪羊

开我东阁门
坐我西阁床

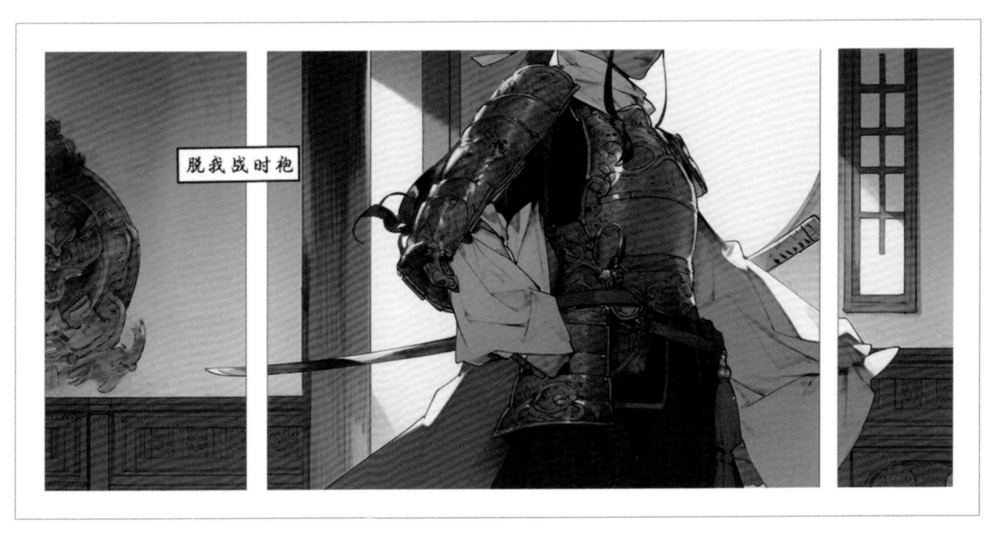
脱我战时袍

著我旧时裳

当窗理云鬓

对镜帖花黄

出门看火伴

火伴皆惊忙

同行十二年

B078- 李泽钦、E182- 陶醉、E194- 狐说、H366- 崔国彬、H391- 小六一 AI

关键词：

风格： in style of naraka: bladepoint（按照游戏《永劫无间》的风格）
主体： 描述主体行为
画风： ink painting（水墨画）
配色可修改： bold color scheme（大胆的配色方案）
更多细节： realistic detail, super mass, 32k, 2d（逼真细节，超高质量，32K，2D）
灯光： zhang yimou's movie lighting effects（张艺谋的电影灯光效果）
后缀： super detail --ar 16:9 --niji 5 --style expressive --q 2--s 750（超级细节 --ar 16:9--niji 5--style expressive --q 2--s 750）
整体： in style of naraka: bladepoint, 描述主体行为 , ink painting, bold color scheme, realistic detail, super mass, 32k, 2d, zhang yimou's movie lighting effects, super detail --ar 16:9 --niji 5 --style expressive --q 2 --s 750（按照游戏《永劫无间》的风格，描述主体行为，水墨画，大胆的配色方案，逼真细节，超高质量，32K，2D，张艺谋电影灯光效果，超级细节 --ar 16:9 --niji 5 --style expressive --q 2 --s 750）

4. 大圣回家

关键词：

by wu guanzhong, warrior character, full body, flat color illustration, sun wukong, the oblique side of the monkey king, standing at high point overlooking mountains --ar 1:1 --v 5

作者吴冠中，武侠人物，全身，平面彩色插图，孙悟空，美猴王的斜侧面，站在高处俯瞰群山 --ar 1:1 --v 5

关键词：

by wu guanzhong, warrior character, full body, flat color illustration, a lot of people on the clouds, a lot of gods on the clouds, all kinds of gods, all gods, people crowded, no empty space, the sky, the white clouds --ar 12:6 --v 5

作者吴冠中，武侠人物，全身，平面彩色插图，云上有很多人，云上有很多神仙，各种神仙，所有神仙，人群拥挤，没有空隙，天空，白云 --ar 12:6 --v 5

关键词：

by wu guanzhong, warrior character, full body, flat color illustration, a thousand people eating at a table in the square, square table, full of people, no space, no white space, crowded, everybody's huddled together, everybody's next to everybody else --ar 12:9 --v 5

作者吴冠中，武侠角色，全身，平面彩色插图，一千人在广场的桌子上吃饭，方桌，满满是人，没有空间，没有空白，拥挤的，大家都挤在一起，人挨着人 --ar 12:9 --v 5

关键词：

by wu guanzhong, one mountain after another, trees on the mountains, birds in the sky, very quiet --ar 12:9 --v 5

作者吴冠中，一座一座的山，满山的树，天上有鸟，非常安静 --ar 12:9 --v 5

关键词：

by wu guanzhong, warrior character, full body, flat color illustration, sun wukong, the back of the monkey king, the back view, a thousand people climbing a mountain, with the huge mountain taishan in the background, everybody's huddled together, everybody's next to everybody else --ar 12:9 --v 5

作者吴冠中，武侠角色，全身，平面色彩插图，孙悟空，美猴王的背面，后视图，一千人在登山，背景是巍峨的泰山，人人挤作一团，人挨着人 --ar 12:9 --v 5

1423 – 王志鹏

关键词：

by wu guanzhong, warrior character, flat color illustration, sun wukong, monkey king, facial close-up, happy, longing expression --v 5

作者吴冠中，武侠人物，平面彩色插图，孙悟空，美猴王，面部特写，快乐，渴望的表情 --v 5